SPACE-TIME INFORMATION PROCESSING

Alan Winder & Charles J. Loda

WESTPORT, CONNECTICUT USA

SPACE-TIME INFORMATION PROCESSING

Reprint Edition Published by:
Peninsula Publishing
Westport, Connecticut, USA

E-mail: sales@PeninsulaPublishing.com
Telephone: 203-292-5621
Website: http://www.PeninsulaPublishing.com

Library of Congress Catalog Number 80-83559

ISBN-10: 0-932146-04-X

ISBN-13: 978-0932146-04-5

Printed in the United States of America

PREFACE TO REPRINT

This book originally appeared under the title "Introduction to Acoustical Space-Time Information Processing". Almost twenty years have passed since Winder and Loda wrote this classic for the Office of Naval Research, and yet the material remains fresh and valuable. In an era concerned with the technology of the fast Fourier transform and digital filtering, it is all too easy to inadvertently neglect the fundamentals with which this work is concerned, namely the limitations in either time or frequency placed on any processing system.

"Space-Time Information Processing" consists of eleven major sections. The first is an introduction and covers the Fourier transform, statistical analysis, correlation and spectra. The next four sections address the various aspects of measuring functions that are spatially and temporally limited. The sixth section is a fine summary of orthogonality and integral transforms. The next three sections form a group dealing with linear systems (such as in circuits), linear spatial systems, and optimal filtering procedures. The last sections address specific acoustical interpretations of results.

This is a grand reference work; it discusses the theory rather than proving it and it backs up its discussions with extensive citations of the literature. These alone are most useful.

There is a wealth of material in "Space-Time Information Processing" and anyone doing signal processing or data analysis will benefit from its use as reference tool. It is a great source for basics and nuances that should have been taught in school but somehow never came up. As an author and consultant in the field of signal processing, I must confess that in only the first twenty minutes of reading, I found two gems of information that were of significant help to me.

Robert K. Otnes, Ph.D.
Palo Alto, California
March 1981

PREFACE

During the past two decades much attention has been directed toward the development, use, and analysis of acoustical information systems for a wide range of applications. The requirements of these systems have become increasingly complex and the research associated with them has produced a wealth of experimental data, instrumentation, and theoretical analyses. Further, advances in circuit system and information theory have provided additional tools for use directly or indirectly in the solution of problems associated with these requirements.

Despite the availability of these theoretical tools and instrumentation, many critical problems remain. The need is therefore evident for improving methods of describing and analyzing complex information processes, both to assist in the selection of measuring instrumentation and in the design of acoustical systems.

Typical problem areas are those associated with sonar. Here there is frequently a need for simultaneously fulfilling a number of functions such as the determination of the existence of a number of signal sources, their localization, isolation in space, and their identification. Analytical methods must describe quantitatively the effectiveness of performing these functions within the limitations imposed by such constraints as power, time, bandwidth, and spatial extent. Additional constraints almost invariably arise from lack of complete knowledge regarding the space-time characteristics of the boundaries and other characteristics of the medium in which the system must operate. These uncertainties must be understood and preferably should be an explicit part of the analysis. Beyond this, logistic factors such as complexity and reliability require consideration. Similar problems arise in radar, optics, radio astronomy, and many other areas, but associations are complicated by complex and specialized terminologies.

This document has been prepared by the authors under Office of Naval Research sponsorship. It presents an introductory study of the interrelations among techniques which are presently available and outlines some of the problem areas for which existing methods are not applicable. Hopefully it will serve as a guide and stimulant for the additional work required to solve the many difficult problems which remain.

Washington, D. C.
October 1962

Aubrey W. Pryce
Director, Acoustics Programs
Office of Naval Research

CONTENTS

CONTENTS (Continued)

CONTENTS (Continued)

CONTENTS (Continued)

INTRODUCTION

This survey attempts to provide guidelines within which analytical and instrumental aspects of multidimensional information-processing may be examined. In all practical cases, a large number of processing procedures are permissible, and a large number are, in fact, necessary. It would be desirable to have rules for determining quantitatively the most economical, or the most effective of the techniques. However, their formulation is not easily or simply done, and discussions of only a few introductory facets comprise this entire volume.

The distinguishing feature of the work is the attempt made to bring together analytical and instrumental tools which appear to be of value in improving the understanding of problems associated with multidimensional, acoustical information processes, particularly those requiring both time and space variables for their descriptions. Complete and unified treatment has not been possible and consequently the survey should be considered as an introduction to problems rather than as a detailed exposition of solutions.

Contemporary literature of communications, detection, and information theory, and the rapid evolution of radar and communication systems reflect the greater effort in electromagnetic areas relative to acoustics. Until recently, point-to-point transfer of information had been the major concern in electromagnetics. The bulk of analytic work is still characterized by having spatial details elided or analyzed separately in noninformational terms. There are many pragmatic reasons for this, since complete space-time representations may become prohibitively complex, and in fact, for some applications such analysis may not be justified. Despite the complexity it is necessary to attempt consolidating basic concepts which are associated with multidimensional processing.

It is important first to recognize some of the factors which have influenced the evolutionary trends of electromagnetic and acoustic information systems such as radar and sonar. In sonar, the relatively low velocity of propagation of energy in conjunction with the use of audio frequencies established requirements for spatially-complex systems such as highly directional, multibeam transmission and reception. These systems were required in order to overcome low data rates and to reduce the effects of the many sources of interference within the medium, including the effects of multiple-path propagation. The use of low frequencies facilitated the early development of components permitting electronic beam-forming of large arrays. In electromagnetics, the propagation velocity provided inherently high data rates and electronic beam forming was not considered an important requirement and in fact was not realizable until recently. The advent of high-speed radar targets imposed increasingly severe functional requirements necessitating extremely large transmitting and receiving arrays, high power, and complex waveforms. Limitations on the performance of many electromagnetic systems - radar and communications - are now no longer solely attributable to stationary noise processes within the receiver, but arise from external interactions with a complex environment. Since these problems have characterized sonar from its earliest days to the present, performance is not to be evaluated simply, in threshold detection terms alone, but involves fulfillment of a number of functions, such as resolving and tracking multiple targets having a wide range of levels. Other features may include distinguishing spatial extent and shape, and temporal changes of these characteristics. In terms of equipment economy, operational necessity, or under environmental conditions where propagation characteristics may be time-varying, many of the operations are to be performed simultaneously or in rapid sequence. Not all of the functions may be performed independently, and determining the nature of constraints comprises an important element in the understanding of physical processes and system analysis. Under the joint influence of technological advances and severe performance requirements many problems within previously unrelated areas such as optics, communications, radio astronomy, radar, and sonar have converged. An interchange of ideas and techniques may now be possible, and, hopefully this document will facilitate such an interchange.

At one time, information theory was regarded as a potential unifier which would provide solutions of problems within several areas in terms common to each of them. Unfortunately, this has not completely materialized since unifying philosophies have an unfortunate tendency to become highly specialized subjects in themselves, thereby compounding rather than easing the problems of interdisciplinary translations. Although this tendency exists, it is not the sole difficulty associated with unification. Despite morphological similarities which appear within many areas, fundamental differences also exist, and a detailed understanding of the physical processes within each area always remains as an important requirement. Analytical models applicable within one area may inadequately represent physical processes in other areas. The inadequacy may arise from assumptions made to simplify analysis, or the models may be so complex that physical properties are obscured. As a result, the evolution of instrumentation, measurement procedures, and of systems may be misdirected, or may be unduly influenced by gadgeteering when analytical models are completely ignored. This survey is intended to reflect a belief that improved understanding of information processing can result from a blending of mathematics which describes real physical processes and contains an attitude toward mathematics and instrumentation which makes them each contributing, rather than controlling, components of the blend.

A brief review of the organization which has been followed will be presented. Because of the strong interrelationship which exists among the topics of the survey it is difficult to develop their significance in a satisfying, logical sequency. The major topics include analytical descriptions of the structural details of functions and physical elements, discussions of information processes, and an introduction to system analysis. Representations of functions may not be detached from element analysis, which in turn, is strongly associated with information and decision theory. Emphasis on the analysis of components tends to obscure system considerations wherein interrelationships - including the order in which components are assembled - may have an importance equal to or greater than the characteristics of components considered in isolation.

Initially, analytical representations of functions are discussed illustrating some of the many methods which are available for representing structural detail of functions. Of importance are the dual requirements for matching descriptions to the characteristics of the functions, including bounds which are imposed, while taking into account the use which is to be made of the representation. The intent may be to improve understanding or visualization of a problem, to facilitate computations, or to make the ultimate realization of instrumentation more economical. No single class of representations can be expected to fulfill each of these requirements optimally. Consequently, a large number of methods have been developed. Rigorous and exact procedures cannot be formulated for the selection of a method for a particular problem, but there are certain invariant features which, when recognized, serve as useful guidelines.

The methods of Fourier analysis serve as an important introduction. Such methods are regarded from the beginning not merely as procedures associated with harmonic analysis, but as basic techniques for the transformation of variables. Representations, sampling, and transformations are seen to be related operations with an important common characteristic being a conversion of structural detail with the intent to simplify functions or to reduce the complexity of operations on the functions. Not only analysis is to be simplified - there may be a preferred domain or structure for physical realization and measurement. Measures of completeness, of conservation, or invariance must be applied, since simplification is not a sufficient criterion for quantitative analysis. Since two or more domains may be involved, the correlation of errors of analytical approximations, and of measurement must be known or established among them. Fourier representations inherently involve mean-square-error as a criterion, and when the structural components represent amplitudes of physical variables, then measures of completeness are in energy terms. Properties of Fourier transforms and the use of Fourier related descriptions for deterministic and random processes are outlined. These include autocorrelation functions, power spectra, probability density functions, and characteristic functions. The use of statistical descriptions may arise in several ways. The only information available or obtainable may be statistical, or the complexity associated with the use of deterministic descriptions may be such that statistical analysis involving a reduction of dimensionality is a practical necessity. Additional descriptions involving joint relationships are required as the number of independent variables increases - these may include crosscorrelation functions, cross-power spectra, joint probability density functions, and joint characteristic functions. The concepts of linear and statistical independence, coherence, and of orthogonality are discussed. Properties of gaussian distributions are seen to have special significance in multidimensional problems.

Although time is implicitly and explicitly involved as an independent variable in the preliminary discussions, the same procedures may be extended to include spatial problems. Spatial sampling concepts apply directly in one or two dimensions. Representation of a radiated field associated with discrete radiators may be made with Fourier series with the element excitation acting as the coefficients in the series. For continuous radiators the field can be expressed as the Fourier transform of the amplitude distribution. The resulting equations are valid at large distances from the radiating aperture and when the distribution does not vary too rapidly in intensity in terms of the wavelength of radiation.

Procedures related to the representation of functions are used for descriptions of physical elements. Earlier it was indicated that the two topics are closely related. A specific example of this relationship is the description of time-bounded functions in which the characteristics of the function generator are used as an integral part of the representation. Element analysis requires determining the relationship between the input and output. Specific uses of the superposition integral and system function - being time and frequency structural representations of elements - are given. Such descriptions are related through the Fourier transform. Element analysis and synthesis include not only measures of completeness, but also of physical realizability and stability. Choice of the representation may be influenced by additional considerations required when multiple elements are involved.

Related problems are discussed which arise in the analysis of spatial elements. Optical imaging elements may be treated as two-dimensional space-frequency filters and techniques of circuit theory may be applied to their analysis. Descriptions corresponding to the impulse response and transfer function are the point source response and the transfer function. The transfer function describes the contrast reduction for a series of sinusoidal patterns of increasing spatial frequency. Here, too, one domain may be preferred for analysis or measurement. In optics, it is often more convenient to measure the reduction in contrast of a periodic test object than to determine the light distribution in an image.

Although similar concepts may be employed in antennae design, greater complexity arises in conjunction with the availability of the spatial dimensions and time as an additional variable. Normally, reflector antennae and two-dimensional arrays may be regarded as space filters and their analysis may be effected by linear, time-invariant network theory. However, if use is to be made of the additional dimensions, then the analysis is more complex. The equivalent network is time-varying with the output being a function of the modulation of the antenna parameters and.the input signal. However, multiplicity of spatial patterns will be available, each pattern comprising, in a sense, an independent information channel. The greater dimensionality available in array design may also involve operations in the near and far fields, both in transmission and reception. Additional degrees-of-freedom may be obtained from the use of polarized radiation. However, their utility may be determined only after careful analysis has established the nature of the dependencies within the various domains.

In order to illustrate some of the types of constraints which may arise in representations of structure a few basic examples are discussed in detail. It is important to recognize that, in general, the number and type of structural components are not intrinsic properties of a particular problem but are actually functions dependent on the mode of representing or defining the problem. Earlier it was stated that imposing bounds on functions had influence on the type of representation which could be used. For example, when a common point of origin in time does not exist, phase information may not be significant, and descriptions such as power spectra, autocorrelation functions, or low-order statistical moments may be used. If the process is bounded, however, important restrictions may arise. Concepts as instantaneous power spectrum and spectral correlation are reviewed. Spectral correlation, for example, permits distinguishing between a random function and one which is switched on and off periodically. Bounds imposed jointly in conjugate domains involve additional constraints, and the descriptions, for example, of time and bandlimited functions is found to be dependent on the manner in which the effective "occupancy" in the time and frequency domains has been defined.

For some problems it is desirable to have the product of the time and frequency occupancy a minimum. Spectrum analyses of transient signals, and simultaneous measurements of frequency and time of arrival of pulsed signals, are required in many problems. A common characteristic of such descriptions and operations is that an indeterminancy exists which imposes a limit on the number of independent structural components. One method of describing the indeterminancy is by a joint autocorrelation function. This function indicates that although the distributions within the joint domains may be altered by nonsimultaneous, that is, sequential operations, the structure defined by this function is invariant to combined displacements in time and frequency. Similar relationships occur in spatial problems involving bounds on aperture distributions and the angular spectra associated with the radiated energy. Bounds imposed jointly set limits on the resolving power and rejection of an antenna or lens. Other indeterminate relationships exist, for example, the formation of an image in an optical system, or a reflector antenna is a function of time. In order to establish the steady-state image or directivity pattern, the length of a pulse must be at least as long as the aperture. Since the pulse length represents a measure of range resolution there exists a limiting value of the combined angular and range resolution. Increasing aperture size to improve angular resolutions may decrease range resolution. It is of interest to note that human performance exhibits similar characteristics when subjected to multiple tasks. The span of absolute judgement is a term applied to the description of the structural detail which may be performed. Different limiting values or spans result depending on whether the challenge is in a single mode or in two or more modes simultaneously. In quantum mechanics, a number of "uncertainty relationships" have been defined. These give the limits associated with simultaneous operations on canonically conjugate variables. Recognition of the relationship has provided valuable conceptual and quantitative guides in modern physics. Studies of constraints associated with multiple, simultaneous operations associated with Fourier-related variables are playing an increasingly important roll in information processing.

Preceding discussions are concerned with descriptions of structural aspects only. The one concept which was stressed involved determining the number of degrees-of-freedom which may be required - or which may be available and, up to this point, explicit. Considerations of the physical environment have not been necessary since structural detail may be specified a priori. It is this factor which permits formulating and solving problems in communications, optics, radar, sonar and radio astronomy by analogous methods. It is also indicated that some understanding of the physical processes is required, particularly in spatial problems since a direct correspondence does not exist among the several areas. The full importance of detailed understanding of the physical processes arises when the totality of information is analyzed - including not only structure but also the range of observable values which can be associated with the structure. There are a number of representations of information, and their complete review would transcend the scope of this work and hence only a few basic facets are discussed.

Structural representations were discussed without reference to the disturbing influence of noise or experimental errors. Measures of completeness were seen to be related to the presence of bounds, including those jointly imposed on conjugate variables. The number of intervals which can be observed or measured within a structural representation, however, will be limited by thermal noise occurring at some stage of the observation process. The total information will consist of the number of discernible points within the complete structure - which may include space and time variables. This measure constitutes the total number of steps required for identifying or selecting a representation from an ensemble of possible representations. It also constitutes a measure of the information obtained from a measurement. Measurements are characterized by a number of limitations which always put a finite bound on information. Infinitely fine structural detail cannot be physically observed because bounds limit resolution, and the detail within the structure is limited to finite values because of the unavoidable presence of errors or noise. Structural "noise" may also occur, for example, in conjunction with time-varying propagation parameters. Important aspects of informational processing require determining information content, information rate, or information density, relating these measures to the fulfillment of specific functions, and establishing measures of cost or efficiency. The total, complete evaluation is dependent on the nature of the bounds imposed on the problem, and on the characteristics of the physical environment. For multidimensional problems, complex interrelations may exist and decision theory rules are required to establish quantitative assessments.

In order to illustrate the use of these concepts a number of informational processes are discussed. These include filtering operations with the desired and undesired information assuming a variety of structures. Problems associated with multiple, matched filtering and analogies in spatial processes are discussed. Combined space-time operations are also outlined. These problems have been selected from a number of areas in order to illustrate similarities and differences within acoustics and electromagnetics. An introduction to system problems within the several areas is also made with emphasis on the relationships of information sources, the propagating medium, and receivers. Finally, a summary review is presented for acoustics and electromagnetics using the major topics of the survey as the organizational elements.

A large amount of detail is presented; a vastly greater amount has been omitted. Although it is difficult to present generalized evaluations of significance, there are some features which deserve reemphasis. The ultimate use of information processing involves making decisions based on measurements of physical processes. The processes of interest in this survey are characterized by large dimensionality - specifically involving spatial and temporal variables. Information processing requires sequences of many transformations from the physical to the decision environment. The transformations are made to effect simplification and may involve reduction of dimensionality, or matching. Quantitative measures are required to establish their effectiveness, and in order to simplify analysis it may be necessary to apply several criteria in the analysis of complex system problems. These criteria may include energy transfer, information conservation, or a number of statistical decision rules. An important key to multidimensional analysis involves determining the degree and nature of dependencies which constrain the effectiveness of the processing. The existence of bounds such as finite time, spatial extent, bandwidth, energy, and the presence of noise all combine to limit the total information. Representations of structural detail, since these details are specifiable *a priori*, may be made by analogous methods for many acoustic and electromagnetic processes permitting an interchange of analytical and instrumental techniques in areas such as optics, circuit theory, radar, radio astronomy, and sonar. However, for complete analysis it is necessary to consider the totality of information which involves detailed *a posteriori* knowledge of the environment, and to provide a quantitative assessment of effectiveness and costs associated with the processing.

Unfortunately it is necessary to consider specific system problems individually. However, some elements of the philosophy contained may be useful in minimizing time and effort spent searching for improvements of components when performance may be inherently constrained by the bounds which have been imposed on the problem. Under such conditions the philosophy outlined in the survey would suggest a diversion or search for other domains where additional degrees of freedom may be obtained. The synthesis of large antennal structures by the motion of simple elements as has been performed by radio astronomers and in surveillance radar comprise excellent examples of the concept and of the constraints on such operations. Numerous other examples are discussed in the text. It is evident that some of the fundamental differences between acoustical and electromagnetic processes which exist may be used to advantage in information processing through the combined use and interaction of acoustical and electromagnetic energy.

The individual elements described by this survey are not original. The significance of the work consists in its organization, that is, in the attempt made to trace systematically basic mathematical and physical concepts which characterize space-time information processing. Many of the actual descriptions are in themselves quite familiar. In many instances, individual descriptions and illustrations have been taken directly from reference works and exact and complete acknowledgment may not have been made in all instances. The document may contribute little to specialists or to the fortunate individuals who have already developed their own philosophy of information processing. Hopefully it may be of some value as a guide to individuals who lack the considerable time which is required to develop their own philosophy by wading through the voluminous literature currently available.

The following references have been helpful in establishing the basic concepts which have been described in the introduction.

1. R. B. Lindsay, H. Margenau, "Foundations of Physics," Chapters 1, 2, 3, John Wiley and Sons, 1936

2. H. Margenau, "The Nature of Physical Reality," Chapters 5, 7, 9, 12, 18, 20, McGraw-Hill, 1950

3. C. G. Darwin, "Observation and Interpretation," pp. 209-18, from book of same title edited by S. Korner, Butterworths Scientific Publication, 1957

4. C. H. Page, "Physical Mathematics," D. Van Nostrand, 1958

5. E. P. Wigner, "The Unreasonable Effectiveness of Mathematics in the Natural Sciences": Comm. on Pure and Applied Mathematics, Vol. XIII No. 1, February 1960

6. L. Brillouin, "Science and Information Theory," Chapters 8, 12, 14, 15, 16, 20, Academic Press, 1956

7. L. Brillouin, "Mathematics, Physics, and Information," Information and Control Vol. 1, No. 1, September 1957

8. R. Vallee, "A Note on Algebra and Macroscopic Observation," Information and Control Vol. 1, No. 1, September 1957

9. Y. Bar-Hillel, "An Examination of Information Theory," Phil. Sci. Vol. 22, pp. 86-105 (1955)

10. D. Gabor, "Communication Theory and Physics," I.R.E. Inf. Theory Trans., February 1953

11. D. M. MacKay, "Quantal Aspects of Scientific Information," Philosophical Magazine, Vol. 41 Ser. 7, No. 314, March 1950

12. J. L. van Soest, "Some Consequences of the Finiteness of Information," Proc. of Symposium on Inf. Theory, Butterworths Scientific Publications (1955)

13. H. Quastler, "Studies of Human Channel Capacity," Proc. of Symposium on Int. Theory, Butterworths Scientific Publications (1955)

14. S. Goldman, "Information Theory," Chapter 9, Prentice Hall, 1953

15. J. Rothstein, "Information, Organization, and Systems," Inf. Theory Trans., September 1954

16. W. H. Huggins, "Representation and Analysis of Signals," Part VII, Signal Detection in a Noisy World, AFCRC-TN-60-360, September 1960

17. W. H. Huggins, "Signal Theory," IRE Trans. on Circuit Theory Vol. CT-3, No. 4, December 1956

18. E. E. David, Jr., "Signal Theory in Speech Transmission," IRE Trans. on Circuit Theory, Vol. CT-3, No. 4, December 1956

19. C. H. Page, "Applications of the Fourier Integral in Physical Science," IRE Trans. on Circuit Theory, Vol. CT-2, No. 3, September 1955

20. R. M. Lerner, "Design of Signals," Lectures in Communication System Theory, McGraw Hill Book Co., 1961

21. R. M. Lerner, "Representation of Signals," Lectures in Communication System Theory, McGraw Hill Book Co., 1961

22. W. E. Morrow, Jr., "Channel Characterization: Basic Approach," Lectures in Communication System Theory, McGraw Hill Book Co., 1961

23. W. M. Siebert, "Signals in Linear Time-Invariant Systems," Lectures in Communication System Theory, McGraw Hill Book Co., 1961

24. D. G. Brennan, "Probability Theory in Communication System Engineering," Lectures in Communication System Theory, McGraw Hill Book Co., 1961

25. D. A. Bell, "Information Theory and its Engineering Applications," Chapters 2, 3, 8, Sir Isaac Pitman and Sons, 1957

26. P. M. Woodward, "Probability and Information Theory with Application to Radar," Pergamon Press (1957)

27. D. McLachlan, Jr., "Description Mechanics," Information and Control, Vol. 1, No. 3, September 1958

28. R. Madden, "The Indeterminancies of Measurements Using Pulses of Coherent Electro-Magnetic Energy," PIEE, Part C, March 1961

29. F. P. Adler, "Minimum Energy Cost of Observation," Inf. Theory Trans., Vol. 1T-1, September 1953

30. D. M. MacKay, "The Structural Information Capacity of Optical Instruments," Information and Control, Vol. 1, No. 2, May 1958

31. F. J. Zucker, "Summary Comments," Proceedings of Symposium on Communication Theory and Antenna Design, Jan. 1957

32. P. Elias, "Optical Systems as Communications Channels," Proc. of Symposium on Information Networks, April 1954

33. J. S. Burgess, "The Future of Radar," Trans. of Milit. Elec., April 1961

34. W. M. Siebert, "A Radar Detection Philosophy," IRE Trans. on Inf. Theory, Vol. IT-2, September 1956

35. R. N. Goss, "A Survey of the Detection Problem," USNEL Research Report 734, 18 Jan. 1957

36. J. L. Stewart and E. C. Westerfield, "A Theory of Active Sonar Detection," PIRE, May 1959

37. J. Ide, "Development of Underwater Acoustic Arrays for Passive Detection of Sound Sources," PIRE, May 1959

38. E. Eichler, "Limitations of Angular Radar Resolution," Proc. of Cont. of Milit. Elect., 1960

39. H. E. Shanks and W. R. Bickmore, "Four-Dimensional Electromagnetic Radiators," Can. Journal of Physics, Vol. 37, 1959

40. H. Gano, "An Aspect of Information Theory in Optics," 1960 Conv. Record, Part 4, Auto. Cont., Inf. Theory

41. H. Wolter, "On the Application of the Basic Theorem of Information Theory to Optics," Physica, Vol. 24, No. 6, pp. 457-75 (1958)

42. G. Toraldo Di Francia, "Capacity of an Optical Channel in the Presence of Noise," Opt. Acta Vol. 2, No. 1, April 1955

43. G. Toraldo Di Francia, "Supergain Antennas and Optical Resolving Power," Suppl. Nuovo Cimento, Vol. 9, pp. 426-438 (1952)

44. E. L. O'Neill, "Spatial Filtering in Optics," IRE Trans. on Inf. Theory, June 1956

45. O. J. M. Smith, "Mixed Distributed and Lumped Parameter Systems," WEscon Conv. Rec., Part 2, 1957

A. REPRESENTATIONS OF FUNCTIONS

1. DETERMINISTIC ANALYSIS

INTRODUCTION

The intent of the initial section of the survey is to discuss various analytical methods for representing some of the functions which are important in information processing. It is neither possible nor desirable to present completely the many facets, and only a few aspects which illustrate basic attitudes will be discussed.

Two requirements are involved when representing structural detail of functions. The description must "match" the characteristics of the function and the use which is to be made of the representation. The latter may involve improving understanding or visualization of a problem, facilitating computations, or making the ultimate realization of instrumentation more economical. No single class of representations may be expected to fulfill each of these optimally, and a large number of methods have been developed. Although rigorous procedures cannot be formulated for the selection of a method for a particular problem, some useful guidelines exist.

Fourier analysis constitutes an important introduction to the methods of representing functions, not only historically, but as a basic and useful technique. It is to be regarded as more than harmonic analysis, incorporating the fundamental concepts of a transformation of variables — that is, the conversion of structural detail with the intent to simplify functions or to reduce the complexity of operations on the functions. The simplification may involve not only analytical operations but there may be a preferred domain or structure for measurement or physical realization. Since simplification is not an adequate criterion for quantitative analysis, measures of completeness must be applied. Fourier representations inherently involve mean-square-error as a criterion, where the structural components represent amplitudes of physical variables, and the measure is in energy terms.

FOURIER SERIES

The Fourier series constitutes an excellent introduction to representations that are not bounded, that is, those that do not have fixed origin or epoch and hence are invariant under a displacement of time, and which involve linearity. Linearity requires that added causes produce added effects independent of the effects of previous causes. The analysis must support the properties of both invariance and linearity. A linear analysis into trigonometric terms is an example.

The simplest representation of a function by linearily-additive trigonometric terms is that provided by the theory of Fourier series. A Fourier series may be used to represent an arbitrary function $f(t)$ over a time interval T if $f(t)$ is absolutely integrable over the interval, i.e.,

$$\int_0^T |f(t)|dt < \infty \qquad \text{(A-1)}$$

and, if the total rise plus the total fall of the function in the interval is finite. This permits applying a minimization (or maxima) criterion with respect to some characteristic of the original function in order to evaluate the completeness of the representation. Alternate specifications which permit analysis by a Fourier series require the function $f(t)$ to contain a finite amount of energy in the interval T (integrable square), i.e.,

$$\int_0^T |f(t)|^2 \, dt < \infty . \tag{A-2}$$

The value of the function at a point of discontinuity is assumed to be the average of the right and left limiting values (continuous in the mean), i.e.,

$$f(t)|_{t=t_0} = \frac{1}{2}\left[\underset{\epsilon \to 0}{\text{limit}}\; f\left(t_0 + |\epsilon|\right) + \underset{\epsilon \to 0}{\text{limit}}\; f\left(t_0 - |\epsilon|\right)\right] . \tag{A-3}$$

A class of functions that meets these restrictions can be expressed in trigonometric form as

$$f(t) = \frac{A_0}{2} + \sum_{n=1}^{\infty}\left[A_n \cos \frac{2\pi}{T} nt + B_n \sin \frac{2\pi}{T} nt\right] . \tag{A-4}$$

This is known as the Fourier series where A_n and B_n are the Fourier coefficients

$$A_n = \frac{2}{T}\int_{-T/2}^{T/2} f(t) \cos \frac{2\pi}{T} nt \, dt \; ; \qquad B_n = \frac{2}{T}\int_{-T/2}^{T/2} f(t) \sin \frac{2\pi}{T} nt \, dt . \tag{A-5}$$

It is possible to represent the Fourier series in terms of complex exponentials,

$$f(t) = \sum_{n=-\infty}^{\infty} D_n e^{j\frac{2\pi}{T}nt} \tag{A-6}$$

where D_n are the complex Fourier coefficients and D_{-n} their complex conjugates

$$D_n = \frac{1}{T}\int_{-T/2}^{T/2} f(t) e^{-j\frac{2\pi}{T}nt} \, dt . \tag{A-7}$$

Positive and negative harmonics, or positive and negative frequencies, are considered equally; two conjugate complex coefficients are furnished for the frequency terms and their sum represents the real coefficients given by (A-4).

Completeness of the Fourier series representation may be considered by examining a finite number of terms in (A-6) and the error function ϵ_N

$$\epsilon_N = \int_{-T/2}^{T/2} \left| f(t) - \sum_{-N}^{N} D_n e^{jn\frac{2\pi}{T}t} \right|^2 dt . \tag{A-8}$$

Use of the error function establishes an integral-square error criterion and gives the interpolation error in terms of energy. By expanding (A-8) a set of orthogonal coefficients D_n is determined which make ϵ_N a minimum. When this is done the Fourier coefficients given by (A-7) are obtained and thus the Fourier series is an orthogonal representation. The integral-square error criterion is an implicit property of the Fourier series. The error vanishes as the number of terms in the expansion becomes infinite.

An important theorem in Fourier series analysis is Parseval's theorem. This states that when D_n is given by (A-7), the error function (A-8) will tend to zero in the limit as $N \to \infty$ and

$$\frac{1}{T}\int_{-T/2}^{T/2} |f(t)|^2 \, dt = \sum_{-\infty}^{\infty} |D_n|^2 . \tag{A-9}$$

Hence, if the $f(t)$ represents a physical process, such as a pressure-time function, the average energy associated with the function is equal to the sum of the average energy in its Fourier components.

If the coefficients of a harmonic series expansion are chosen to be Fourier coefficients, then the integral-square error will have its smallest possible value. From Parseval's theorem, a finite number of Fourier terms will be a better approximation of the original function with respect to energy content than a similar number of terms of any other orthogonal representation. It is this aspect which makes Fourier expansion such a useful tool, for it is this and only this expansion which shows how the energy is distributed in frequency. Figure A-1 illustrates how the sum of the first three terms of the Fourier Series expansion of a train of pulses approximates the function.

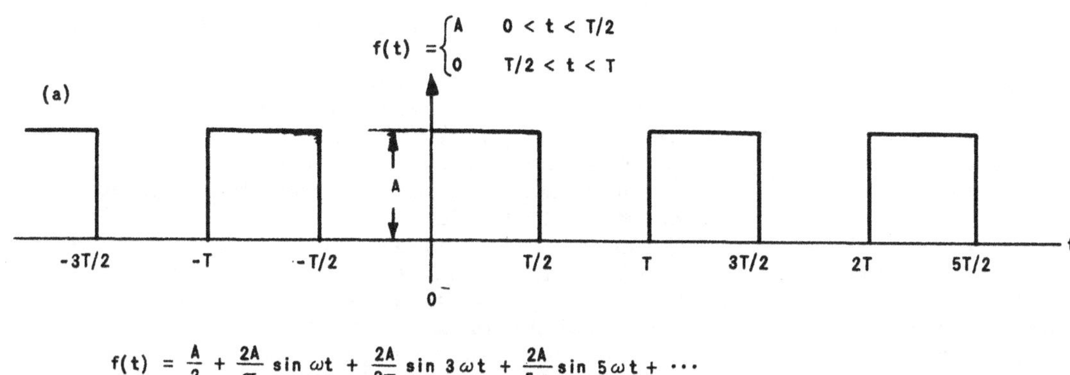

$$f(t) = \frac{A}{2} + \frac{2A}{\pi}\sin\omega t + \frac{2A}{3\pi}\sin 3\omega t + \frac{2A}{5\pi}\sin 5\omega t + \cdots$$

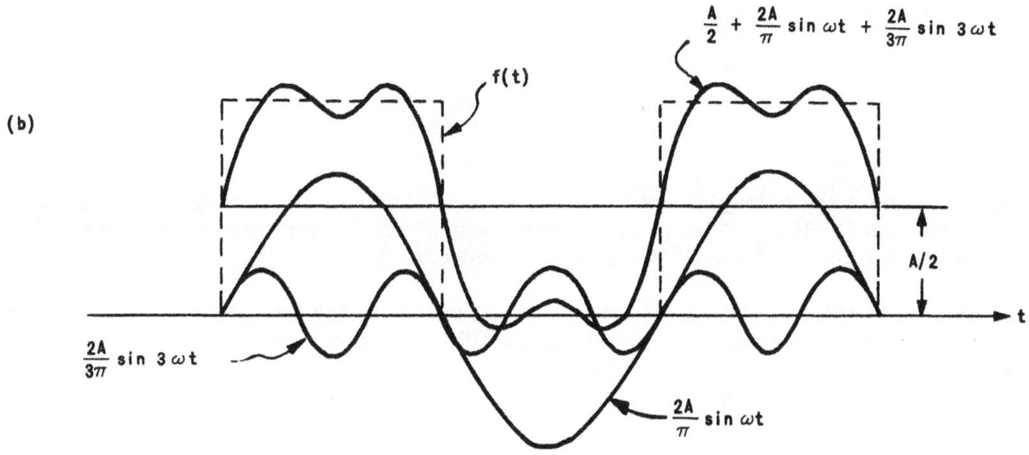

Figure A-1 - Illustration of howthe sum of the first three terms of the Fourier series expansion of a train of pulses approximates the function

LINE SPECTRA

Two spectra can always be obtained from a Fourier analysis. A complete and practical characterization of a function can be given by a graph with the harmonic number n as abscissa and the Fourier coefficients A_n and B_n as ordinates. They are called the Fourier cosine-series spectrum and the Fourier sine-series spectrum according to whether the A_n or B_n coefficients are plotted. Since n is a discontinuous variable, each spectrum will consist of a set of discrete lines.

The line spectrum, using form (A-6), can also be used to characterize the amplitudes and phase angles of the harmonic expansion at the harmonic frequencies. The amplitude and phase spectra are even and odd symmetrical functions of frequency, respectively.

CONTINUOUS SPECTRUM

The concept of a spectrum can be extended to include noncontinuous or transient functions. In the discussion of the Fourier series, reference to a periodic function was deliberately omitted since the series may also be used for the representation of functions within a given interval. It is important to note that line spectra as representations are confined to periodic functions. If the function is nonperiodic, the phase and amplitude spectra will be continuous.

The transition from a line to a continuous spectrum is illustrated in Figure A-2 where the continuous spectrum of a pulse is obtained by letting the period of an infinite train of pulses, Figure A-2(a), become infinite. Figure A-2(b) shows that when the period is doubled with the pulse height and width being unchanged, the zero frequency component (the average value of the wave) is halved and the spectrum will contain more harmonics but have the same envelope. In Figure A-2(c), the period is again doubled and the results are similar to those in Figure A-1(b). As the period is increased indefinitely, the continuous spectrum of an isolated pulse is finally obtained as shown in Figure A-1(d). In the limit, the value of the zero frequency component and the spacing between the harmonic frequency components becomes zero. This continuous spectrum may be regarded as consisting of an infinite number of "components." The magnitudes of the components are infinitesimal and cannot be measured by direct graphical methods. Calculus techniques over finite time intervals must be used.

FOURIER RELATIONSHIP BETWEEN THE WAVEFORM AND SPECTRUM

The continuous counterpart to the complex Fourier series representation of a waveform given by (A-6) is

$$f(t) = \frac{1}{2\pi} \int_{-\infty}^{\infty} F(j\omega)\, e^{j\omega t}\, d\omega \tag{A-10}$$

where $F(j\omega)$ is the amplitude-phase spectrum, $\omega = 2\pi f$ is the radian frequency, a continuous variable. Equation (A-10) is valid provided the integral exists, and indicates that the Fourier coefficients are a discrete form of the Fourier spectrum.

The waveform $f(t)$ and its amplitude-phase spectrum $F(j\omega)$ form a Fourier transform pair. That is, when the waveform is given by (A-10), the Fourier spectrum may be expressed as

$$F(j\omega) = \int_{-\infty}^{\infty} f(t)\, e^{-j\omega t}\, dt\ . \tag{A-11}$$

The Fourier spectrum is determined by the complete history of the waveform from $t = -\infty$ to $t = +\infty$. When it is expressed in terms of frequency f, the complex notation on the left-hand side of (A-11) is usually omitted, and $F(j\omega)$ becomes $F(f)$.

A useful theorem in Fourier transform analysis is Plancherels' theorem. It states that if $F(f)$ is of integrable square on the entire range, $-\infty < f < \infty$, then there is a function $f(t)$ which is also of integrable square on the entire range, $-\infty < t < \infty$. The functions are related by the following:

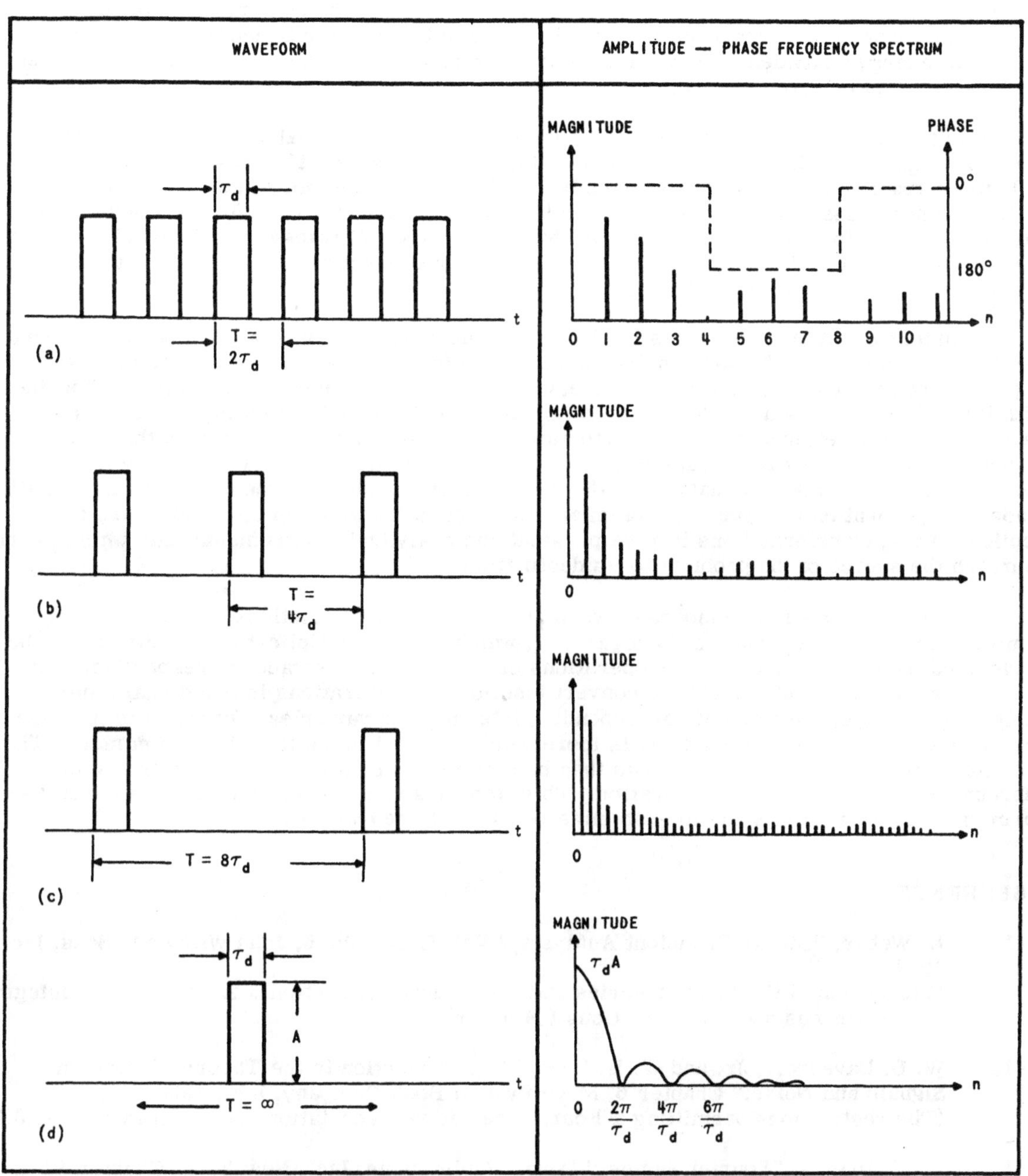

Figure A-2 - Transition of the Fourier line spectrum of a train of pulses to the continuous spectrum of an isolated pulse

$$\int_{-\infty}^{\infty} |F(f)|^2 \, df = \int_{-\infty}^{\infty} |f(t)|^2 \, dt . \qquad \text{(A-12)}$$

Since $|f(t)|^2$ is the instantaneous power, (A-12) represents the total energy of the waveform and $|F(f)|^2$ is the energy spectrum. Plancherels' theorem is the continuous analogue of Parseval's theorem and measures the completeness of the Fourier transformation by establishing an energy equivalence between the waveform and its spectrum. Similarly, it reduces the integral-square error to zero.

Despite the limitations and difficulties associated with analytical and practical aspects of Fourier transforms, the benefits to be derived are considerable. The most powerful aspects of Fourier transform theory may perhaps be attributed to extensibility, that is, the inherent ability to encompass a wide range of physical processes. Fourier transform theory historically has set the stage for other transforms. A characteristic of all transforms derivable from the Fourier transform is that they possess a Plancherel type theorem and consequently, an integral-square error for evaluation purposes.

Transforms and the procedures implied, "transformations," have always been fundamental tools. One definition of "transform" is "to change in form, shape, nature, function, as an algebraic expression or geometrical figure, without altering the meaning or value." Together, with this, should be added the definition of its synonym, "convert" which is, "a change of the details which are better suited for a particular use." These definitions contain the essence of transform theory. That is, a transform is a tool that provides a greater flexibility for the application of mathematical analysis to the reduction of a problem. The type of transformation made is dependent on the type of problem and the application to which the results are to be applied. Thus, transformations involve physical and analytical "instrumentation" which intend to match the source of the problem to its destination.

Transforms pertain to numbers or variables, as well as to functions of variables. A familiar "number transform" is the logarithm, which allows multiplication and division to be performed by means of the simpler operations of addition and subtraction, respectively. A function transform should be able to convert functions and operations in one domain into simpler (algebraic) functions of corresponding intermediate variables. Then, as in the case of the logarithm, the algebraic equation is more readily solved in the transformed domain. The solution in the original domain(s) could then be obtained by consulting the appropriate tables, thereby performing an inverse operation. This, too, is another advantage in using the transform method, for it allows one to systemize results obtained with it.

REFERENCES

A-1.1. E. Weber, "Linear Transient Analysis," Vol. I, Chapter 6, John Wiley and Sons, Inc., 1957
(Discussion of the Fourier series and line spectra in section 6.2. and Fourier integral and continuous spectra in sections 6.4 to 6.6)

A-1.2. W. B. Davenport, Jr. and W. L. Root, "An Introduction to the Theory of Random Signals and Noise," Chapter 6, McGraw-Hill Book Company, Inc., 1958
(The restrictions permitting a Fourier series representation are given in section 6.1)

A-1.3. W. E. Byerly, "Fourier Series," Dover Publications, Inc., New York, Chapter VI
(The convergence of a Fourier series and Dirichlet's conditions are discussed in sections 33 and 37, respectively)

A-1.4. R. Courant and D. Hilbert, "Methods of Mathematical Physics," Volume I, Chapter II, Interscience Publishers, Inc., 1953
(General discussion of the expansion of an arbitrary function in an orthogonal series)

A-1.5. C. R. Wylie, Jr., "Advanced Engineering Mathematics," Chapter 5, McGraw-Hill Book Company, Inc., 1951
(General discussion of the Fourier series representation)

A-1.6. E. C. Tichmarch, "Introduction to the Theory of Fourier Integrals," Oxford University Press, London, 1937

A-1.7. E. A. Guillemin, "The Mathematics of Circuit Analysis," Chapter VII, John Wiley and Sons, Inc., 1956
(General treatment of Fourier analysis)

A-1.8. P. Franklin, "Fourier Methods," McGraw-Hill, New York, 1949

A-1.9. W. L. Root and T. S. Pitcher, "On the Fourier Series Expansion of Random Functions," Annals of Math. Statistics, Vol. 26, No. 2, pp. 313-318, June 1955

A-1.10. R. Paley and N. Wiener, "Fourier Transforms in the Complex Domain," Am-Math. Soc. Colloquium Pubs., Vol. XIX, Am-Math. Society, New York, 1934

2. STATISTICAL ANALYSIS

INTRODUCTION

The need for applying probabilistic or statistical methods may arise in a number of ways. As a result of the influence of many variables which may not be readily measured, a deterministic specification of the problem may prove impractical from both a physical and analytical point of view. "Noise," in general, fits in this category, since it is not possible to predict on the basis of a measurement made at one region in space, at one instant of time, what the precise value of a noise voltage or current will be at a future time or another spatial region. However, given certain bounds on the process, useful estimates of future values or of values in other spatial regions may be possible. In other cases, the influences of the variables may be known or may be determinable, in principle. However, the complexity of a deterministic description could be so great as to affect its utility, and statistical descriptions may consequently be favored because of their relative simplicity.

Statistical descriptions are based upon assumption of a type of regularity. For example, in order to establish a mathematical model of statistical events, we must assume that as the number of experiments giving rise to the events are repeated without limit, they will tend to some "smoothed" or regular behavior. This process is termed statistical regularity and may give rise to a discrete or continuous random variable(s), though in most cases it is both.

An associated step in determining a suitable description is to establish a quantitative measure of the variable. This is done through the notion of a probability value. A positive number is assigned to a particular event which behaves as the limit of the relative frequency of occurrence of that event with respect to the total number of events as the latter becomes infinite. We can then speak, given a random process, of the probability of obtaining a particular result from that process. For a random process, the random variable varies not only with respect to its position in some "space" but also as a function of time.

A set of functions possessing one or more characteristic properties, such as a collection of sine waves, is called an "ensemble" of functions if the set has a probability distribution given with it. In fact, to affect a statistical analysis, our state of knowledge must be the probability distribution function of the random variables. We can then make statistical predictions of future values of a function of the ensemble. The stochastic or random variables are also called the degrees-of-freedom of the ensemble and are the number of values needed to specify the function at any one instant of time.

DISCRETE AND CONTINUOUS PROBABILITY DENSITY FUNCTION

A function of a random variable equal to the probability of obtaining the random variable as it goes through a whole range of values is called a probability distribution function. Although the probability distribution function offers a complete statistical description of a random process, in most problems it cannot be determined easily by direct means, and other descriptions are required. One of these is the probability density function which is the derivative of the probability distribution function with respect to the random variables involved. In order to make this definition applicable to discrete random variables the concept of impulse functions is employed. If all possible values of the random variables are considered as describing a "field," then the probability density function of any point in the "field" is proportional to the probability of finding the random variables in a differential region containing the point.

If $P(x \leq X)$ is the probability distribution function of a random variable x, where X is any value in the range of x, then the probability density function $p(x)$ is given as:

$$p(X) = \frac{dP(x \leq X)}{dX} \tag{A-13}$$

such that

$$P(x \leq X) = \int_{-\infty}^{X} p(x)dx \,. \tag{A-14}$$

In differential notation, (A-13) may be expressed as

$$P(X-dX < x \leq X) = p(X)dX. \tag{A-15}$$

The probability that a random variable x falls in the interval $a < x \leq b$ is the difference between the values of the probability distribution function at the end points of the interval, i.e.,

$$P(a < x \leq b) = P(x \leq b) - P(x \leq a) \,. \tag{A-16}$$

If $x(t)$ is a random function, the probability of finding $x(t)$ at a particular value, X, for a given time is zero. Instead, the probability of finding $x(t)$ within the interval X and $X - dX$, is defined. From (A-15), this probability may be expressed in terms of the probability density function $p(X)$ and the differential interval dX as $p(X)dX$. The probability of finding $x(t)$ equal to any value between $x = a$ and $x = b$ would be the sum of the probabilities of finding it within any one of the strips that make up the interval (if the probabilities are mutually exclusive, that is, if they do not occur together). For a continuous random function, this may be written as:

$$P(a < x \leq b) = \int_{a}^{b} p(x)dx \,. \tag{A-17}$$

Since it is certain that (x) will be found somewhere, using (A-14),

$$\int_{-\infty}^{\infty} p(x)dx = 1 \,. \tag{A-18}$$

The probability density function of a discrete random variable may be considered to be comprised of an impulse at each of its possible values having a strength equal to the corresponding probability.

$$p(x) = \sum_{m=1}^{M} P(x_m)\,\delta(x - x_m) \tag{A-19}$$

is the probability density function for a discrete random variable which has M possible values x_m with probabilities $P(x_m)$. The function $\delta(x - x_m)$ is the impulse function centered upon $x = x_m$.

AVERAGES

The usefulness of a probability density function is that from it may be derived averages which are better adapted for analysis and measurement. These include the mean value, the mean-square value, and averages of higher positive powers of random functions.

The concept of a statistical average involves the limit of the arithmetical average of a random variable. If x is a discrete random variable taking on any one of M possible mutually exclusive values x_m, the statistical average $E[g(x)]$ of a single-valued function of x, $g(x)$, which would also be a discrete random variable is defined by the equation

$$E[g(x)] = \sum_{m=1}^{M} g(x_m)P(x_m) \tag{A-20}$$

where $P(x_m)$ is the probability of occurrence of value x_m.

If x is a continuous random variable with probability density function $p(x)$, the statistical average of the continuous random variable $g(x)$ is defined as

$$E[g(x)] = \int_{-\infty}^{\infty} g(x)p(x)dx\,. \tag{A-21}$$

Equation (A-21) may be extended to the case where x is a mixed discrete and continuous random variable by allowing $p(x)$ to contain impulse functions. It is important to note that the statistical average of a function of a given random process may be a function of time.

The statistical average of the nth power of random variable x is called the nth moment, m_n of its probability density function and is given as

$$m_n = E\,[x^n] = \int_{-\infty}^{\infty} x^n\,p(x)dx. \tag{A-22}$$

For $n = 1$ and $n = 2$, the corresponding moments are equal to the mean value and mean-square value, respectively. In electrical terminology, m_1 represents the dc component of the process and m_2 gives the mean power dissipated in a one-ohm resistor. It is often found convenient to deal with the ac or systematic components only. The averages are then called central moments, μ_n, and are defined by

$$\mu_n = E\,[(x - m_1)^n] = \int_{-\infty}^{\infty} (x - m_1)^n\,p(x)dx\,. \tag{A-23}$$

An important central moment is μ_2, which is defined as the variance of the density function. From definition (A-23) the variance may be put in the form

$$\mu_2 = E[x^2] - (E[x])^2. \tag{A-24}$$

The square root of the variance is defined as the standard deviation σ, i.e.,

$$\sigma = (\mu_2)^{1/2}. \tag{A-25}$$

In electrical terminology, the standard deviation is the rms value of the ac component. The variance is the mean-square value and when multiplied by the conductance or resistance, whichever is appropriate, it gives the mean power represented by the ac component. It is the average ac power dissipated in one-ohm resistor.

The density function is completely determined when moments of all orders exist. Knowledge of the moments is equivalent to a knowledge of the probability density function in the sense that it is possible theoretically to exhibit all properties of the probability density function in terms of moments.

An average that is very important in physical problems is the mean value of the function $\exp(jvx)$, x being a random variable. This is called the characteristic function $C_x(jv)$, and using (A-21), is expressed as

$$C_x(jv) = E[e^{jvx}] = \int_{-\infty}^{\infty} e^{jvx}\, p(x)dx \,. \qquad \text{(A-26)}$$

By comparing (A-26) with the Fourier relationship between a waveform and its spectrum, the characteristic function is the Fourier transform of the probability density function. $p(x)$ may be obtained by taking the inverse Fourier transform of $C_x(jv)$,

$$p(x) = \frac{1}{2\pi}\int_{-\infty}^{\infty} C_x(jv)\, e^{-jvx} dv \,. \qquad \text{(A-27)}$$

Since they are Fourier conjugates, both descriptions can provide equivalent information. The use of one or the other is dependent on the problem. One of the primary attributes of this relationship is that it may facilitate computation of the probability density function by transforming an n-fold integration in the "density function domain" to an n-fold multiplication and one integration when the operations are performed in the "characteristic function" domain.

The averages discussed earlier were derived from the probability density function. It is also possible to determine the moments through use of the characteristic function. The moments are calculated employing auxiliary parameters called semi-invariants or cumulants which are derived from a power series expansion of the natural logarithm of the characteristic function. The first semi-invariant is the mean value and the second is the variance.

Averages may be obtained or defined either as operations in time or over sets of functions at an instant of time. The time average of the function $x(t)$ which is a member of the random process $[x(t)]$ is defined as

$$\langle x(t)\rangle = \lim_{T\to\infty} \frac{1}{T}\int_{-T/2}^{T/2} x(t)dt \qquad \text{(A-28)}$$

where T is the interval over which the average is taken. It is independent of time and gives the dc component while eliminating both the nonsystematic and systematic components of the function. The statistical average, unlike the time average, is usually a time function which eliminates the nonsystematic components completely while retaining the systematic components.

The mean-square value is the average value of the square of the function and may be an average over time or an average over the sample functions at a particular instant. The main difference between them is that the first is a number and the second a function of time. It should be recognized that a mean-square value is a measure of the quantity of a function but tells nothing of its behavior except how its mean-square value varies with time. Thus, a variety of functions may have the same mean-square value.

STATIONARITY AND ERGODICITY

A random process may be considered as consisting of an ensemble of functions that can be characterized by a complete set of probability density functions. If none of the probability densities which describe the random process changes with time such that the statistics measured at any two distinct instants of time are the same, the process is said to be stationary "in the strict sense." If the mean value is a constant function of time and the statistical average of the product of the random function at two instants of time does not depend on absolute time but only on the time difference, then the process is said to be stationary "in the wide sense." A random process which is stationary in the strict sense is also stationary in the wide sense.

If each member of a stationary ensemble of random functions that make up some random process is typical of the ensemble as a whole, then the random process is said to be ergodic. The statistics over a long time interval for any one random function are then the same as the statistics over the ensemble of random functions at any one instant. An ergodic process is always stationary but a stationary process can be nonergodic. Analysis can often be simplified by assuming that a process is ergodic. However, it is not ordinarily possible to demonstrate or to prove that the physical process is ergodic, other than by comparing the results of measurements with predictions derived from the analysis.

TWO-DIMENSIONAL PROBABILITY THEORY

The probability density function of a single variable allows determining the relative occurrence of different magnitudes but not of the time interval involved in observing such a set of values. The knowledge of the statistics of a pair of values separated by specified instants of time is of special importance. It is necessary to determine the probability relations concerning two coordinates x and y which may be dependent on each other — that is, specifying the value of one affects the statistics of the other. Such distributions are called bivariate, or joint probability distributions.

Similar to single-variable probability theory, the probability density function of two coordinates, $p(x,y)$ may be defined as the function which when multiplied by the infinitesimal area $dxdy$ gives the probability that the value of the first coordinate is in the range x to $x + dx$ and the value of the second coordinate is simultaneously within the range y to $y + dy$. The probability that x lies between x_1 and x_2 while y lies between y_1 and y_2 is expressed as

$$\text{Prob}\left[x_1 < x < x_2,\ y_1 < y < y_2\right] = \int_{x_1}^{x_2}\int_{y_1}^{y_2} p(x,y)dxdy. \qquad \text{(A-29)}$$

The conditional probability density function expresses the state of knowledge of one variable knowing the probability of occurrence of another variable, and has essentially the same properties as the density function previously discussed. If $p(x,y)$ is the joint probability density function of the variables x and y, and $p(x)$ and $p(y)$ represent their individual densities, then the conditional probability density function $p(x|y)$ is

$$p(x|y) = \frac{p(x,y)}{p(y)} \qquad \text{(A-30)}$$

where the independent variable is y. If x is the independent variable,

$$p(y|x) = \frac{p(x,y)}{p(x)}\ . \qquad \text{(A-31)}$$

Conditional densities are bounded by zero and one and are at least equal to the corresponding joint densities,

$$p(x|y) \geq p(x,y)\ . \qquad \text{(A-32)}$$

Averages may be computed as previously, except that there are now two integrations to perform,

$$E\left[f(x,y)\right] = \int_{-\infty}^{\infty}\int_{-\infty}^{\infty} f(x,y)\, p(x,y)\, dxdy \tag{A-33}$$

where $f(x,y)$ is the two-dimensional function under consideration. In particular, the central moments μ_{jk} are expressed as

$$\mu_{jk} = E\left[(x - m_x)^j\ (y - m_y)^k\right] \tag{A-34}$$

$$= \int_{-\infty}^{\infty}\int_{-\infty}^{\infty} (x - m_x)^j\ (y - m_y)^k\, p(x,y)\, dxdy \tag{A-35}$$

where m_x and m_y are the mean values of the random variables x and y, respectively. The most important central moment is the quantity μ_{11}. This is referred to as the covariance of x and y and is expressed as:

$$\mu_{11} = E\left[(x - m_x)\ (y - m_y)\right] \tag{A-36}$$

$$= E[xy] - E[x]E[y]. \tag{A-37}$$

The covariance is a measure of the linear dependence between two quantities. Zero covariance implies linear independence but not necessarily statistical independence. Statistical independence must be determined from central moments of higher order.

Analogous descriptions may be obtained for problems of higher dimensionality. Added dimensions result in increased complexity in their representations, not only because more variables are involved, but because of added bounds and possible dependencies among the variables. Although Fourier transform theory is applicable in multidimensional problems, it is often necessary to use other methods such as conformal mapping.

STATISTICAL INDEPENDENCE

Previously it had been indicated that covariance is a measure of the linear dependence between two random variables. When the first joint moment $E(xy)$ of random variables x and y factors into the product of their means,

$$E(xy) = E(x)\ E(y) \tag{A-38}$$

then x and y are linearly independent. Two random variables which are statistically independent are also linearly independent. However, linear independence does not imply that the variables are statistically independent, unless they are jointly Gaussian. A necessary and sufficient condition for the statistical independence of two random variables is that their joint moment factors

$$E\left(x^n y^k\right) = E\left(x^n\right) E\left(y^k\right) \tag{A-39}$$

for all positive integral values of n and k. Statistical independence may also be expressed through the characteristic function, i.e., the joint characteristic function of two random variables will factor into the product of their respective characteristic functions

$$M_{x,y}(jv_1, jv_2) = M_x(jv_1)\ M_y(jv_2)\,. \tag{A-40}$$

Equations (A-39) and (A-40) are equivalent measures.

The concept of statistical independence is of considerable importance, particularly in multidimensional problems, since it indicates the absence of interactions among the variables and permits simpler descriptions of processes. The key to effective analysis of multidimensional problems involves determining the effect various operations have on statistical independence. Analytically, if independence exists, then the n-dimensional joint characteristic function for n random variables is equal to the n-fold product of their individual characteristic functions. An important theorem in statistics is the central limit theorem. This states that the probability distribution of the sum of an indefinitely large number of independent quantities will approach the Gaussian distribution, regardless of the individual distributions. The significance of the Gaussian distributions is reflected in it being completely determined by having a knowledge of its second moment and in not having to examine moments higher than the first to determine independence. However, it is necessary to establish that transformations performed preserve the Gaussian properties.

A measure of the similarity in phase between functions is termed coherence. When two functions $A(t)$ and $B(t)$ are superimposed, the resultant average power in the time interval T, is

$$P_{av} = \frac{1}{T}\int_0^T \left[A(t) + B(t)\right]^2 dt \tag{A-41}$$

$$= \frac{1}{T}\int_0^T A^2(t)dt + \frac{1}{T}\int_0^T B^2(t)dt + \frac{2}{T}\int_0^T A(t)\,B(t)\,dt\,. \tag{A-42}$$

The first two integrals on the right of the above equation represent the average power of the functions taken separately, while the third integral represents an interaction of the two functions that is dependent on their relative phase. If the third integral is zero, the functions are said to be orthogonal. If the integral is equal to $(2/T)$ times

$$\int_0^T |A(t)|\,|B(t)|\,dt\,,$$

the functions are completely coherent. All incoherent functions are orthogonal, but not all orthogonal functions are necessarily incoherent.

It can generally be stated that the resultant average power in the superposition of a number of functions is equal to the sum of the average powers of the individual functions plus twice the sum of the average values of the products of the incoherent components, each product taken with the proper sign.

REFERENCES

A-2.1. W. B. Davenport, Jr. and W. L. Root, "An Introduction to the Theory of Random Signals and Noise," Chapter 4, McGraw-Hill Book Company, Inc., 1958
(Discusses probability density functions, averages, and statistical independence)

A-2.2. P. M. Woodward, "A Mathematical Description of Random Noise," I.R.E. Journal, October 1948
(Discusses statistical averages)

A-2.3. W. R. Bennett, "Methods of Solving Noise Problems," I.R.E. Proc., Vol. 44, No. 5, pp. 609-638, May 1956
(Discusses averages, stationarity and ergodicity, and two-dimensional probability theory)

A-2.4. H. Cramer, "Mathematical Methods of Statistics," Princeton University Press, Princeton, N. J., 1956

A-2.5. W. Feller, "Probability Theory and It's Applications," John Wiley, New York, 1950

A-2.6. H. J. Laning, Jr. and R. H. Battin, "Random Processes in Automatic Control," McGraw-Hill Book Co., New York, 1956

A-2.7. A. N. Kolmogoroff, "Foundations of the Theory of Probability," Chelsea Publishing Co., Inc., New York, 1950

A-2.8. M. Loeve, "Probability Theory," D. Van Nostrand Co., Inc., New York, 1955

A-2.9. J. L. Doob, "Stochastic Processes," John Wiley and Sons, Inc., New York, 1953

A-2.10. A. I. Khinchin, "Statistical Mechanics," Dover Press, New York, 1949

3. CORRELATION AND SPECTRAL ANALYSIS

INTRODUCTION

The analysis employed in signal representation depends on our state of knowledge of the function. If the state of knowledge permits us to predict with probability one the function's exact value at future instants of time, then a deterministic analysis is desirable. However, if our state of knowledge is the probability distribution function, not equal to unity, then we may affect a statistical analysis. Both analyses provide complete descriptions independent of the wave-shape. The first for an arbitrary function over a certain time interval, and the second for an ensemble of functions at particular instants of time.

An analysis which depends on the state of knowledge being the value of functions at two arbitrary instants of time is called a correlation and spectral analysis and may be obtained through time or statistical averaging. Correlation and spectral descriptions are derivable from a deterministic or statistical analysis but do not provide complete descriptions, except for Gaussian processes. Rather, they offer a qualitative and quantitative measure of trading time-space for frequency-space or time-space for some nonconjugate space in signal representation. Correlation and spectral analysis also helps distinguish temporal aspects in statistical problems. In fact, for most practical problems it will yield identical results whether determined temporally or statistically.

AUTOCORRELATION FUNCTION

One method of representing the dependence between the values of an arbitrary function at two specified instants of time is by averaging their product, either temporally or statistically. This type of description is called an autocorrelation function.

In the case of a statistical representation observations are made at the instants t_1 and t_2 on a large number of similar random functions, thereby obtaining an ensemble of paired values. From (A-33), the autocorrelation function $\psi(t_1, t_2)$ of a random function $x(t)$, may be written as

$$\psi(t_1, t_2) = E\left[x(t_1)\, x(t_2)\right] = \int_{-\infty}^{\infty}\int_{-\infty}^{\infty} x_1 x_2 p(x_1, x_2)\, dx_1 dx_2 \tag{A-43}$$

where x_1 and x_2 refer to the values of the function at instants t_1 and t_2 respectively, and $p(x_1, x_2)$ is the joint probability density function. Comparing (A-43) with (A-36), the autocorrelation function determined statistically is equivalent to the covariance, except for a bias factor. If the random process is stationary, its statistics and consequently, autocorrelation function, will not depend on the particular values of t_1 and t_2 but only on the time difference $\tau = t_2 - t_1$.

In the case of temporal representation, the autocorrelation function of a member of a random process is defined as

$$\psi(\tau) = \underset{T \to \infty}{\text{limit}} \frac{1}{2T} \int_{-T}^{T} x(t)\, x(t+\tau)\, dt. \tag{A-44}$$

In general, (A-43) and)A-44) will not yield identical results. However, if the random process is ergodic, the ensemble statistics and time statistics coincide, and the autocorrelation functions obtained are equivalent. In fact, this can be used as a definition of ergodicity.

It may be seen from (A-44) that the autocorrelation function for $\tau = 0$ is the mean-square value of $x(t)$, while its value for a random process for $\tau \to \infty$ is the square of the mean value. $\psi(\tau)$ is usually a damped function of τ and while it does not define a process uniquely (unless it is Gaussian), it can provide an indication of its "time-constant." The autocorrelation function also has the properties

$$\begin{aligned} \psi(\tau) &= \psi(-\tau) \\ |\psi(\tau)| &\leq \psi(0). \end{aligned} \tag{A-45}$$

These features of the autocorrelation function are illustrated in Figure A-3.

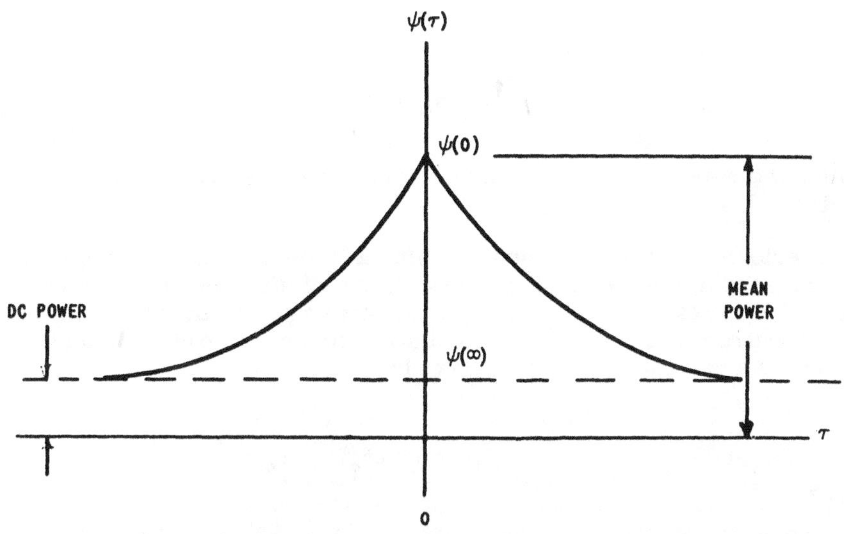

Figure A-3 - General autocorrelation function

The definition (A-44) may be applied to an arbitrary function of time as well as to a sample function of a random process so long as the indicated limit exists. If the function $x(t)$ is periodic and can be represented by a Fourier series, i.e.,

$$x(t) = \sum_{n=-\infty}^{\infty} D_n e^{jn\omega_o t} \tag{A-6}$$

then the autocorrelation function using (A-44) is found to be

$$\psi(\tau) = D_o^2 + 2 \sum_{n=1}^{\infty} |D_n|^2 \cos n\omega_o \tau . \tag{A-46}$$

The autocorrelation function of a periodic function is comprised of its dc value squared plus all of its harmonics. Notice should also be taken of the absence of all phase angles. All periodic time functions which have the same Fourier coefficient magnitudes and periodicities also have the same autocorrelation function even though the phases of their Fourier components (and hence their actual time structures) may be different. This indicates that there is a "many-to-one" correspondence between time functions and autocorrelation functions.

The autocorrelation functions discussed previously have been associated with waveforms having infinite total energy content, that is, continuous periodic or random functions. Study of such waveforms, unbounded in either time or frequency serves primarily to facilitate understanding of the mathematical analyses available. Functions that contain a finite amount of energy, that is, energy-bounded functions, are of greater practical significance. Their autocorrelation functions are actually simpler because there is no difficulty with limits,

$$\psi(\tau) = \int_{-\infty}^{\infty} x(t)\, x(t+\tau)\, dt. \tag{A-47}$$

POWER SPECTRUM

If η watts is the average amount of power dissipated in a one-ohm resistance and if the portion of η arising from the components having frequencies between f and $f + df$ is denoted by $W(f)df$, then

$$\eta = \int_{-\infty}^{\infty} W(f)df = \langle x^2(t) \rangle \tag{A-48}$$

and $W(f)$ is called the power spectrum of $x(t)$. $W(f)$ is the spectrum of the average power and has the dimensions of energy.

If $x(t)$ is a periodic function of period T having a finite amount of energy per period, using Parseval's theorem, the power spectrum consists of a series of impulses at the component frequencies of $x(t)$, each impulse having a strength equal to the power in that component. Thus, the power spectrum is a measure of the distribution of the power in $x(t)$ as a function of frequency, and for a periodic function is given by

$$W(f) = \sum_{n=-\infty}^{\infty} |D_n|^2 \delta(f - nf_o) \qquad f_o = \frac{1}{T} \tag{A-49}$$

where D_n is the complex Fourier coefficient given by (A-7). The total power in $x(t)$ is

$$\int_{-\infty}^{\infty} W(f)df = \sum_{n=-\infty}^{\infty} |D_n|^2 = \frac{1}{T}\int_{0}^{T} |x^2(t)|\, dt\,. \tag{A-50}$$

When dealing with a random process comprised of an ensemble of functions the power spectrum $W(f)$ may be characterized by statistical variation from member to member. While it may be completely determinate for any one member of the ensemble, it cannot be plotted in the limit $T \to \infty$, since adjacent values $W(f)$ and $W(f+df)$ will be uncorrelated. It is necessary to reduce these random variations and "smooth" the spectrum from one frequency to another. The resulting description is a "mean-power spectrum." In order to affect a Fourier analysis of a continuous random function $x(t)$ which would have infinite energy content, it is necessary to define an auxiliary function $x_T(t)$ as

$$\begin{aligned} x_T(t) &= x(t) \qquad \text{when } 0 < t < T \\ &= 0 \qquad \text{elsewhere.} \end{aligned} \tag{A-51}$$

The function $x_T(t)$ may now be subjected to Fourier analysis, and as $T \to \infty$, those properties of $x_T(t)$ which approach limiting values will also be properties of $x(t)$. The Fourier transform of $x_T(t)$ is given as

$$X_T(f) = \int_{-\infty}^{\infty} x_T(t)\, e^{-j2\pi ft}\, dt = \int_{0}^{T} x(t)\, e^{-j2\pi ft}\, dt . \tag{A-52}$$

The mean-power spectrum of $x(t)$ is defined as

$$\overline{W(f)} = \lim_{T\to\infty} \frac{2\overline{|X_T(f)|^2}}{T} \tag{A-53}$$

where we consider only values of $f > 0$ and assume that this limit exists. $\overline{W(f)}$, as defined, is a measure of the frequency distribution of the power in the function $x(t)$ which extends from $t = -\infty$ to $t = +\infty$. However, this definition is useful only when $x(t)$ has no dc component or periodic terms. Whenever a random function can be considered as a superposition of disturbances $x(t)$ delayed by varying times so as to form a sequence which is "random in time" with a mean rate λ, then the mean-power spectrum becomes

$$\overline{W(f)} = \lambda 2|X_T(f)|^2 \qquad f \neq 0 . \tag{A-54}$$

The behavior of $\overline{W(f)}$ at $f = 0$ is like an impulse function, for it approaches infinity in such a way as to enclose a finite area.

The information contained in (A-53) or (A-54) as well as in the power spectrum previously defined is less than that in an amplitude-phase spectrum since the phase information has been removed. Any systematic variations are also smoothed out in the derivation of a mean-power spectrum. This is the main advantage of working with the power spectrum. That is, by supressing information about the phase, results are obtained independent of the origin of the time scale. A random function can be represented by its mean-square value, the autocorrelation function or mean-power spectra. Whether these representations are "adequate" will depend on the particular application. In many instances they are used because other representations may not be available or would be too complex. For some, however, they may be entirely adequate. Figure A-4 shows the autocorrelation function and power spectrum for various noise fluctuations having equal noise power.

A fundamental theorem in correlation and spectral analysis is the Wiener-Khintchine Theorem which states that if an arbitrary function is amenable to Fourier analysis, then its autocorrelation function and power spectrum are cosine Fourier transforms of each other. It is this which makes the autocorrelation function such a useful description, that from it we can determine the power spectrum which is often of real interest. The relationship may be written as

$$\psi(\tau) = \int_{0}^{\infty} W(f) \cos 2\pi f\tau \, df \tag{A-55}$$

$$W(f) = 4\int_{0}^{\infty} \psi(\tau) \cos 2\pi f\tau \, d\tau . \tag{A-56}$$

Clearly, if the function is random, we would consider the mean-power spectrum. It should be noted that the Fourier transformations (A-55) and (A-56) are expressed in terms of cosines instead of exponentials because both $W(f)$ and $\psi(\tau)$ are real and even functions.

CROSSCORRELATION FUNCTION

A very important type of correlation is that between two arbitrary functions. This is called the crosscorrelation function and may be obtained by performing a time or statistical

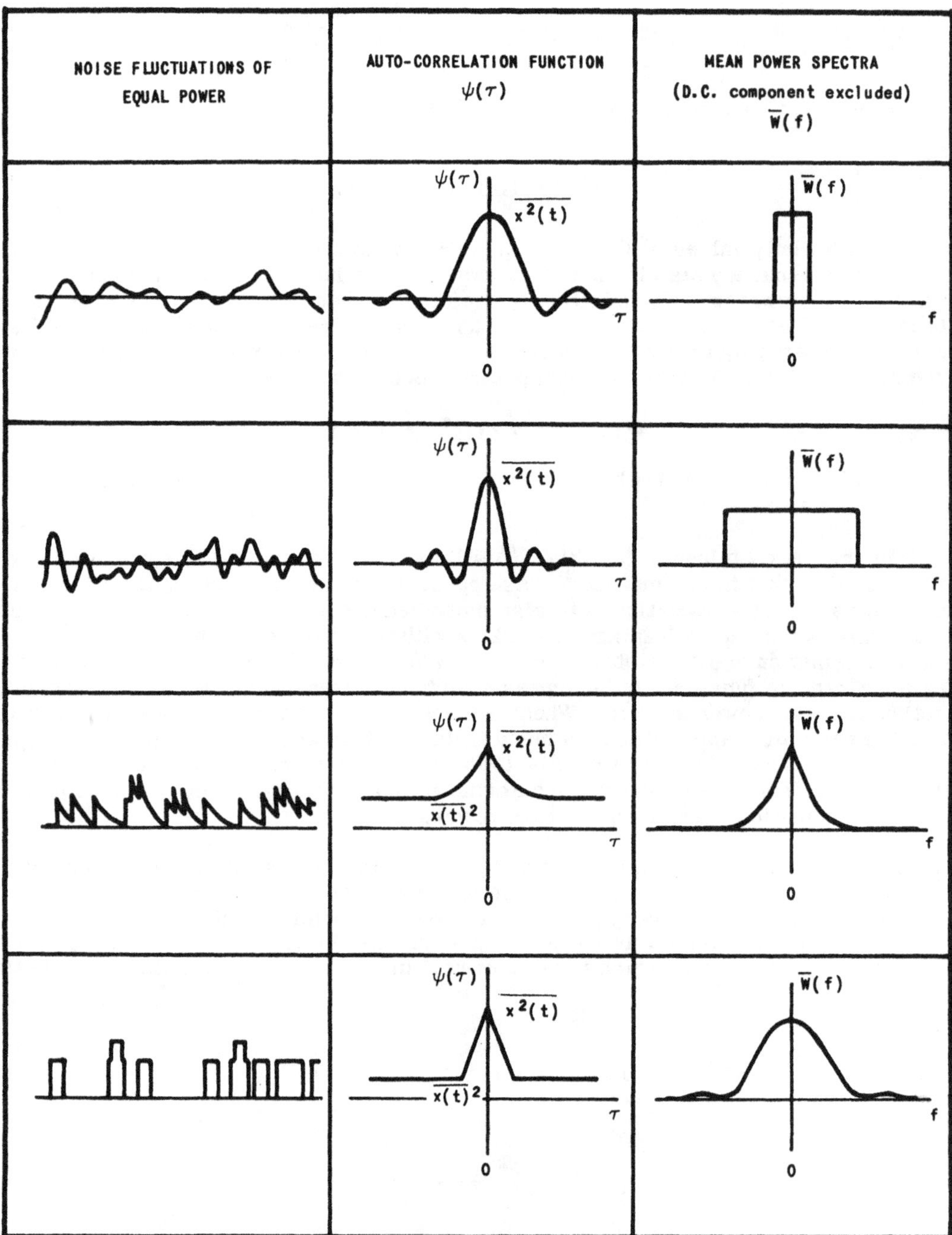

Figure A-4 - Autocorrelation function and mean-power spectra of functions having equal noise power

average of the product of the value of one function at some instant of time with the value of the other function at another instant of time.

For stationary random functions $x(t)$ and $y(t)$ having the joint probability density function $p(x, y)$, the crosscorrelation function ψ_{xy} at the instants t_1 and t_2 is given by

$$\psi_{xy}(\tau = t_2 - t_1) = E\left[x(t_1)\, y(t_2)\right] = \int_{-\infty}^{\infty}\int_{-\infty}^{\infty} x_1 y_2 p(x_1, y_2)\, dx_1\, dy_2 . \tag{A-57}$$

If we average temporally, ψ_{xy} is defined as

$$\psi_{xy}(\tau) = \lim_{T \to \infty} \frac{1}{2T} \int_{-T}^{T} x(t)\, y(t + \tau) dt. \tag{A-58}$$

For ergodic functions, both definitions yield identical results.

If the two functions are periodic having the same fundamental frequency, crosscorrelation retains the fundamental and only those harmonics which are present in both, together with their phase differences. All periodic time functions which have the same Fourier coefficient magnitudes and periodicities with fixed relative phase between the functions will have the same crosscorrelation functions. Thus, similar to the case for autocorrelation, the correspondence between time functions and crosscorrelation functions is a "many-to-one" correspondence. Figure A-5 illustrates the crosscorrelation function of two periodic functions.

When the two functions are incoherent, such as two stationary random functions which are independently generated, crosscorrelation produces a constant, independent of τ, which is a product of the individual mean values of the functions. Although the autocorrelation and crosscorrelation functions are somewhat similar, the crosscorrelation function retains relative phase information, and hence it is necessary to specify whether $x(t)$ or $y(t)$ is taken at the displaced time $t + \tau$. In general,

$$\psi_{yx}(\tau) = \lim_{T \to \infty} \frac{1}{2T} \int_{-\infty}^{\infty} y(t)\, x(t + \tau)\, dt \tag{A-59}$$

need not be the same as $\psi_{xy}(\tau)$. The crosscorrelation function is an even function and has the property

$$\psi_{xy}(\tau) = \psi_{yx}(-\tau) . \tag{A-60}$$

Additionally,

$$|\psi_{xy}(\tau)| \leq \sqrt{\psi_{xx}(0)\, \psi_{yy}(0)} \tag{A-61}$$

where $\psi_{xx}(\tau)$ and $\psi_{yy}(\tau)$ are the autocorrelation functions of $x(t)$ and $y(t)$, respectively.

CROSS-POWER SPECTRA

If $x_1(t)$ and $x_2(t)$ are both zero outside the interval $0 \leq t \leq T$, then the cross-power spectra $W_{12}(f)$ is defined as

$$W_{12}(f) = \lim_{T \to \infty} \frac{X_1(-f)\, X_2(f)}{T} = W_{21}^{*}(f) \tag{A-62}$$

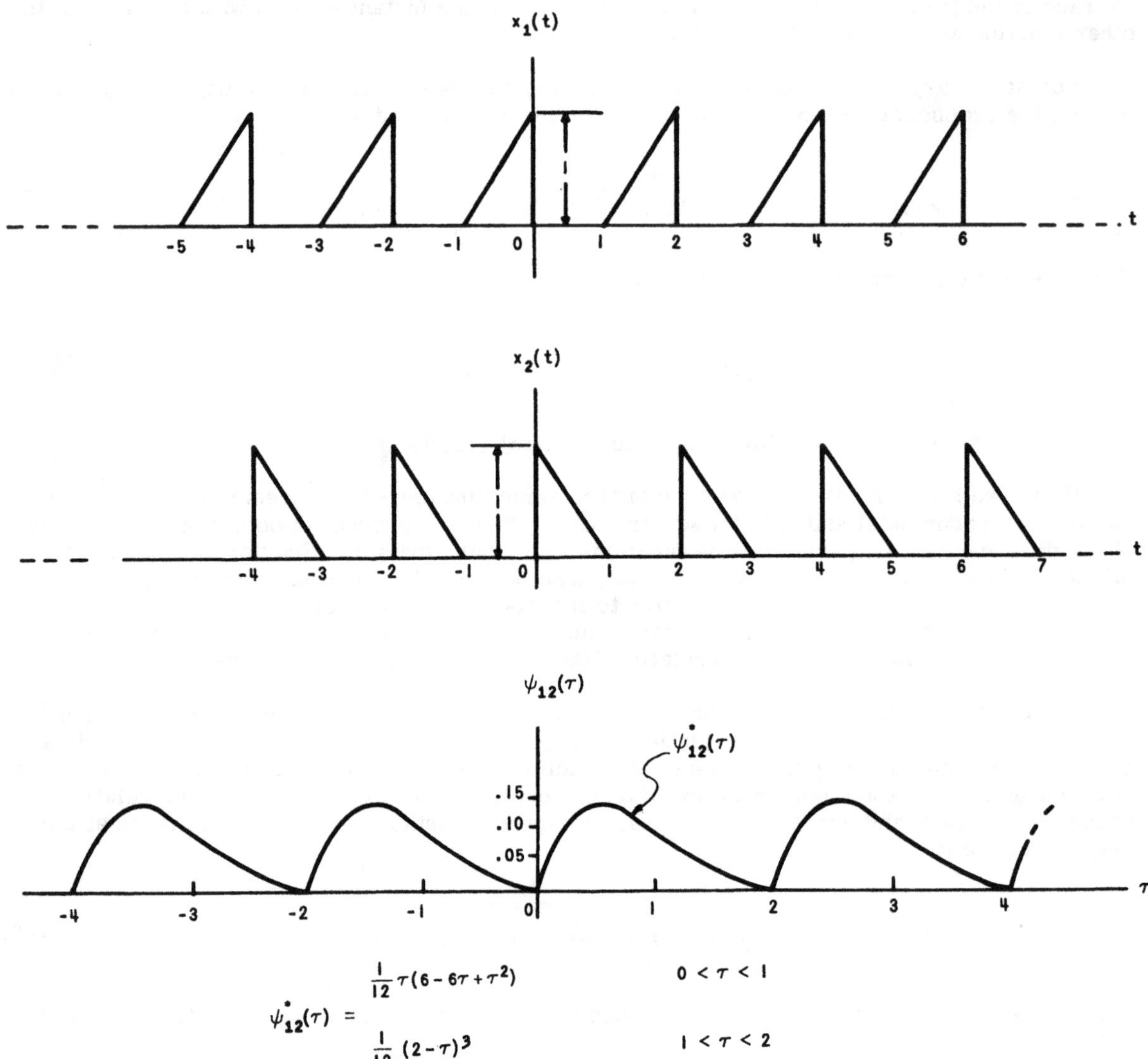

$$\mathring{\psi}_{12}(\tau) = \begin{cases} \frac{1}{12}\tau(6-6\tau+\tau^2) & 0 < \tau < 1 \\ \frac{1}{12}(2-\tau)^3 & 1 < \tau < 2 \end{cases}$$

Figure A-5 - Plot of crosscorrelation function $\psi_{12}(\tau)$ of the periodic functions $x_1(t)$ and $x_2(t)$

where

$$\left.\begin{aligned} X_1(f) &= \int_0^T x_1(t)\, e^{-j\omega t}\, dt \\ X_2(f) &= \int_0^T x_2(t)\, e^{-j\omega t}\, dt \end{aligned}\right\} \qquad \text{(A-63)}$$

The cross-power spectra are seen to be conjugate complex numbers. The real parts are even functions of frequency while the imaginary parts are odd functions of frequency. Therefore, $W_{12}(f)$ may be written as where $U_{12}(f)$ and $V_{12}(f)$ are real and

$$U_{12}(-f) = U_{12}(f) = U_{21}(f) \tag{A-65}$$

$$V_{12}(-f) = -V_{12}(f) = V_{21}(f) . \tag{A-66}$$

If $x_1(t)$ and $x_2(t)$ are incoherent, then $W_{12}(f) = 0$ for all frequencies, although the converse is not necessarily true. If $W_{12}(f)$ does not vanish at some frequency f, the functions are partially coherent. For $|W_{12}(f)|^2 = W_1(f)W_2(f)$ the coherence is total in the second order sense for non-Gaussian sources and totally coherent for jointly Gaussian pairs. If in addition to this condition, $W_{12}(f)$ equals the real quantity $U_{12}(f)$ at all frequencies, then the sources are said to be colinear. Then, for a positive $U_{12}(f)$, the phase will be additive and for a negative $U_{12}(f)$ they will subtract.

If there is total coherence and $W_{12}(f)$ equals the imaginary quantity $jV_{12}(f)$ at all frequencies, the sources are said to be in quadrature, indicating that a 90° phase shift exists between components at the same frequencies. A positive value of $V_{12}(f)$ indicates that the $x_1(t)$ component lags the corresponding $x_2(t)$ component by 90°, while a negative $V_{12}(f)$ indicates a corresponding leading phase angle. Thus, cross-power spectra provide measures of coherence between two arbitrary functions and is shown in Figure A-6 for specific periodic functions.

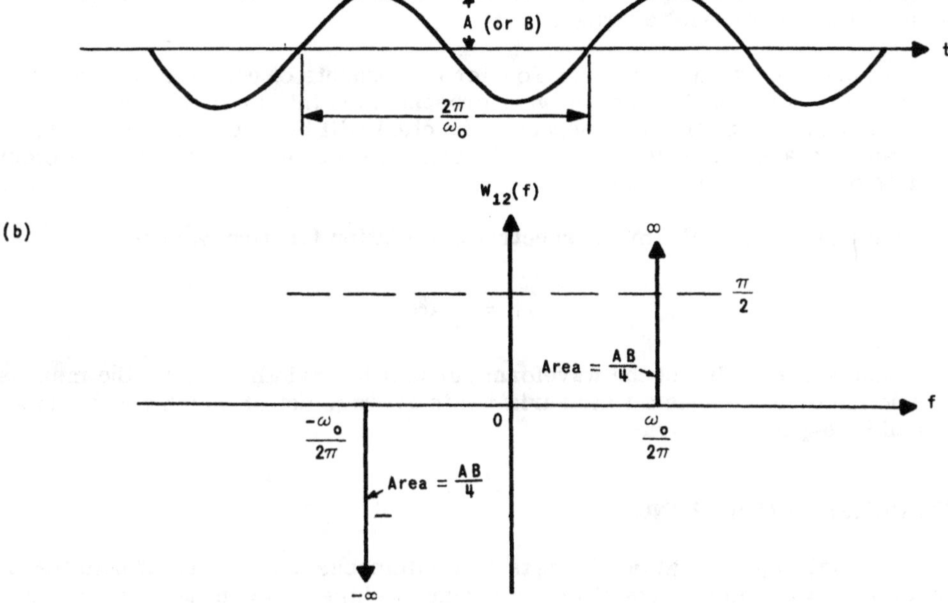

Figure A-6 - Plot of cross-power spectra $W_{12}(f)$ of the periodic functions $x_1(t)$ and $x_2(t)$

The Fourier transform of the crosscorrelation function is the cross-power spectrum. The Fourier transform pair is given by

$$W_{12}(f) = 4 \int_0^\infty \psi_{12}(\tau) \cos 2\pi f\tau \, d\tau \quad \text{(A-67)}$$

$$\psi_{12}(\tau) = \int_0^\infty W_{12}(f) \cos 2\pi f\tau \, df \, . \quad \text{(A-68)}$$

Both functions may be used to describe the degree of coherence between two arbitrary time functions, and either function may be determined from its conjugate mate by performing a Fourier inversion.

MEAN-SPECTRAL CORRELATION FUNCTION

Completely random processes have random amplitude and phase spectra. If the amplitudes are uncorrelated then the power spectrum is discontinuous at all points and the Fourier sine-cosine series have real coefficients which are statistically independent. However, since $W(f)$ does not depend on phase, the power spectrum and the autocorrelation function provide a measure of the correlation between amplitudes.

The mean-spectral correlation function $\overline{\gamma(f)}$ provides a description of noise present in an unvarying amount and noise which is switched repeatedly or "modulated" in some arbitrary manner. This is defined for a given function $x(t)$ as

$$\overline{\gamma(f)} = \frac{1}{B} \int_{-\infty}^{\infty} \overline{X(\phi)\, X^*(\phi - f)}\, d\phi \quad \text{(A-69)}$$

where B is the bandwidth of the spectrum $X(\phi)$ and the asterisk denotes the complex conjugate, and the bar indicates a statistical average.

As indicated previously, the phases of Fourier components of completely random processes are random and uncorrelated. If correlation is present, then this is an indication of something systematic in the process and the mean-spectral function will have a finite value. There are other ways in which phase correlation may be described, such as through higher moments of the probability distribution of the process.

The Fourier transform of the mean-spectral correlation function $\overline{\gamma(f)}$ is

$$\mathfrak{F}\left\{\overline{\gamma(f)}\right\} = \frac{1}{B}\, \overline{x^2(t)} \quad \text{(A-70)}$$

which is the mean-square value of the waveform per unit bandwidth. Hence, the mean-spectral correlation function distinguishes a time variation in average power which results in a correlation of the amplitude-phase spectrum.

JOINT AUTOCORRELATION FUNCTION

A time-frequency representation attempts a simultaneous description of both the time and frequency behavior. Certain aspects of the interrelationships between the conjugate domains can be quantitatively described by the joint time and frequency autocorrelation function. In order to illustrate the significance of a "joint correlation" description, consider the complex signal $f(t) = s(t)e^{j\beta(t)}$ having the Fourier transform $F(f)$, $s(t)$ and $\beta(t)$ representing the variation of amplitude and phase with time, respectively.

If the signal is displaced in time only, the temporal autocorrelation function $\psi(\tau)$ is obtained as follows:

$$f(t) = s(t)e^{j\beta(t)} \tag{A-71}$$

hence,

$$f(t + \tau) = s(t + \tau)e^{j\beta(t+\tau)} \tag{A-72}$$

$\psi(\tau)$ is defined by (A-47) as

$$\psi(\tau) = \int_{-\infty}^{\infty} f(t)\, f^*(t+\tau)dt \tag{A-47.a}$$

(the "delayed" complex conjugate has been introduced to account for the signal being complex.) Substituting (A-71) and (A-72) into (A-47.a),

$$\psi(\tau) = \int_{-\infty}^{\infty} s(t)\, s(t+\tau)e^{-j[\beta(t+\tau)-\beta(t)]}dt. \tag{A-73}$$

If $\beta(t)$ is a linear function of time, say at, (A-73) reduces to the correlation function for the amplitude variation $s(t)$ multiplied by $e^{-ja\tau}$, thus adding a linear phase $\theta = -a\tau$ to the description. This is written as

$$\psi(\tau) = e^{-a\tau} \int_{-\infty}^{\infty} s(t)\ s(t + \tau)\ dt. \tag{A-74}$$

If the signal is displaced in frequency only, it may be described by the phase correlation function, $G(\phi)$, defined as

$$G(\phi) = \int_{-\infty}^{\infty} F^*(f)\,F(f+\phi)df \tag{A-75}$$

where ϕ denotes the shift in frequency. The phase correlation function of a random process may be averaged over the ensemble of functions, with a resulting statistical average phase correlation function that is proportional to the mean-spectral correlation function of the process.

If the signal is displaced both in time and frequency, $f(t)$ becomes

$$f(t + \tau,\phi) = s(t + \tau)e^{j[\beta(t+\tau) + 2\pi\phi(t+\tau)]} \tag{A-76}$$

The complete temporal autocorrelation function $\psi(\tau,\phi)$ is then

$$\psi(\tau,\phi) = \int_{-\infty}^{\infty} f(t)\, f^*(t+\tau,\phi)dt \tag{A-77}$$

Performing the substitutions indicated in (A-77),

$$\psi(\tau,\phi) = \int_{-\infty}^{\infty} s(t)\,e^{j\beta(t)}\,s(t+\tau)\,e^{-j[\beta(t+\tau) + 2\pi\phi(t+\tau)]}\,dt \tag{A-78}$$

$$= \int_{-\infty}^{\infty} s(t)\,s(t+\tau)\,e^{j\beta(t)}\,e^{-j\beta(t+\tau)}\,e^{-j2\pi\phi(t+\tau)}\,dt\,. \tag{A-79}$$

Let $t + \tau = x$ and $dt = dx$. (A-79) then becomes,

$$\psi(\tau,\phi) = \int_{-\infty}^{\infty} s(x-\tau)\, s(x)\, e^{j\beta(x-\tau)}\, e^{-j\beta(x)}\, e^{-j2\pi\phi x}\, dx \tag{A-80}$$

$$= \int_{-\infty}^{\infty} f(x)\, f^*(x+\tau)\, e^{-j2\pi\phi x}\, dx\,. \tag{A-81}$$

Equation (A-81) indicates that the correlation function of a complex signal which undergoes a combined time and frequency shift, denoted by τ and ϕ respectively, may be expressed in terms of waveforms of time and time displacement only, and an exponentially periodic term representing the shift in frequency. The joint autocorrelation function $\psi(\tau,\phi)$ is defined by

$$\psi(\tau,\phi) = \int_{-\infty}^{\infty} f(t)\, f^*(t+\tau)\, e^{-j2\pi\phi t}\, dt \tag{A-82}$$

$$= \int_{-\infty}^{\infty} F^*(f)\, F(f+\phi)\, e^{-j2\pi f\tau}\, df\,. \tag{A-83}$$

A characteristic of the joint-autocorrelation function is that the volume under the surface described by its envelope, $|\psi(\tau,\phi)|^2$, is a constant, i.e.,

$$\int_{-\infty}^{\infty}\int_{-\infty}^{\infty} |\psi(\tau,\phi)|^2\, d\tau\, d\phi = C. \tag{A-84}$$

The constant C is invariant for all waveforms and has been designated as an "absolute structural constant." For a waveform $f(t)$ that is normalized such that

$$\int_{-\infty}^{\infty} |f(t)|^2\, dt = 1 \tag{A-85}$$

then C, the structural constant, is unity.

Similar relationships can be derived for two functions. For two complex waveforms $f_1(t)$ and $f_2(t)$ having Fourier transforms $F_1(f)$ and $F_2(f)$, respectively, the joint crosscorrelation function $\psi_{12}(\tau,\phi)$ is

$$\psi_{12}(\tau,\phi) = \int_{-\infty}^{\infty} f_1(t)\, f_2^*(t+\tau)\, e^{-j2\pi\phi t}\, dt \tag{A-86}$$

$$= \int_{-\infty}^{\infty} F_2^*(f)\, F_1(f+\phi)\, e^{-j2\pi f\tau}\, df\,. \tag{A-87}$$

With the energy in the waveforms normalized to unity,

$$\int_{-\infty}^{\infty} |f_1(t)|^2\, dt = \int_{-\infty}^{\infty} |f_2(t)|^2\, dt = 1 \tag{A-88}$$

the volume under the cross envelope $|\psi_{12}(\tau,\phi)|^2$ becomes

$$\int_{-\infty}^{\infty}\int_{-\infty}^{\infty} |\psi_{12}(\tau,\phi)|^2 d\tau\, d\phi = 1. \quad \text{(A-89)}$$

The joint envelopes of Gaussian pulses are illustrated in Figure A-7. It is seen that if there is no modulation, the major spread will occur along the τ-axis or ϕ-axis, depending on the length of the pulse. However, for linear frequency modulation, the major spread will occur at some angle between the major axis of $|\psi|^2$ and the τ-axis, depending on the complex frequency. A characteristic of linear frequency modulation is that the joint correlation function $\psi(\tau,\phi)$ is not separable, i.e.,

$$\psi(\tau,\phi) \neq \psi(\tau)\, G(\phi). \quad \text{(A-90)}$$

The modulation has introduced a correlation or dependency between the two domains and it appears possible to obtain a wide $|f(t)|^2$ and a wide $|F(f)|^2$ as shown. However, the phase dependency is not shown by $|F(f)|^2$. Consequently, this is not a contradiction of the indeterminate relationship which exists between time duration and bandwidth. Modulation consists in a sense, of a set of sequential operations, and the indeterminency has merely been redistributed.

REFERENCES

A-3.1. W. B. Davenport, Jr. and W. L. Root, "An Introduction to the Theory of Random Signals and Noise," McGraw-Hill Book Co., Inc., 1958
(Discusses the statistical and time correlation functions in sections 4.5 and 4.8, spectral analysis of a wide-sense stationary random process in section 6.6, and cross-spectral densities in section 6.7)

A-3.2. S. O. Rice, "Mathematical Analysis of Random Noise," Part II, Bell System Technical Journal, pp. 23, 282, 1944
(Discusses power spectra and correlation functions)

A-3.3. J. W. Tukey, "The Sampling Theory of Power Spectrum Estimates," Symposium on Applications of Autocorrelation Analysis to Physical Problems, Woods Hole, Mass., pp. 47-67, June 13-14, 1949

A-3.4. P. M. Woodward, "A Mathematical Description of Random Noise," I.R.E. Journal, October 1948
(Discussions of statistical averages, mean-power spectrum, and mean-spectral correlation function)

A-3.5. R. Deutsch, "Detection of Modulated Noise-like Signals," I.R.E. Trans. on Info. Theory, No. 3, pp. 106-122, March 1954

A-3.6. E. Parzen and N. S. Shiren, "Analysis of a General System for the Detection of Amplitude Modulated Noise," Hudson Laboratories Technical Report No. 24, August 1954

A-3.7. C. A. Stutt, "A Note on Invariant Relations for Ambiguity and Distance Functions," I.R.E. Trans., Vol. 1T-5, pp. 164-167, December 1959

A-3.8. P. M. Woodward, "Time and Frequency Uncertainty in Waveform Analysis," Phil. Mag., Vol. 42, pp. 883-891, August 1951

A-3.9. V. V. Solodovnikov, "Introduction to the Statistical Dynamics of Automatic Control," Dover Publications, Inc., New York

A-3.10. M. James, N. Nichols, R. Phillips, "Theory of Servo Mechanisms," Chapter 6, Radiation Lab. Series, Vol. 25, McGraw-Hill Book Co. Inc., New York, 1947

A-3.11. J. L. Lawson and G. E. Uhlenbeck, "Threshold Signals," Radiation Lab. Series, Vol. 24, McGraw-Hill Book Co. Inc., New York, 1950

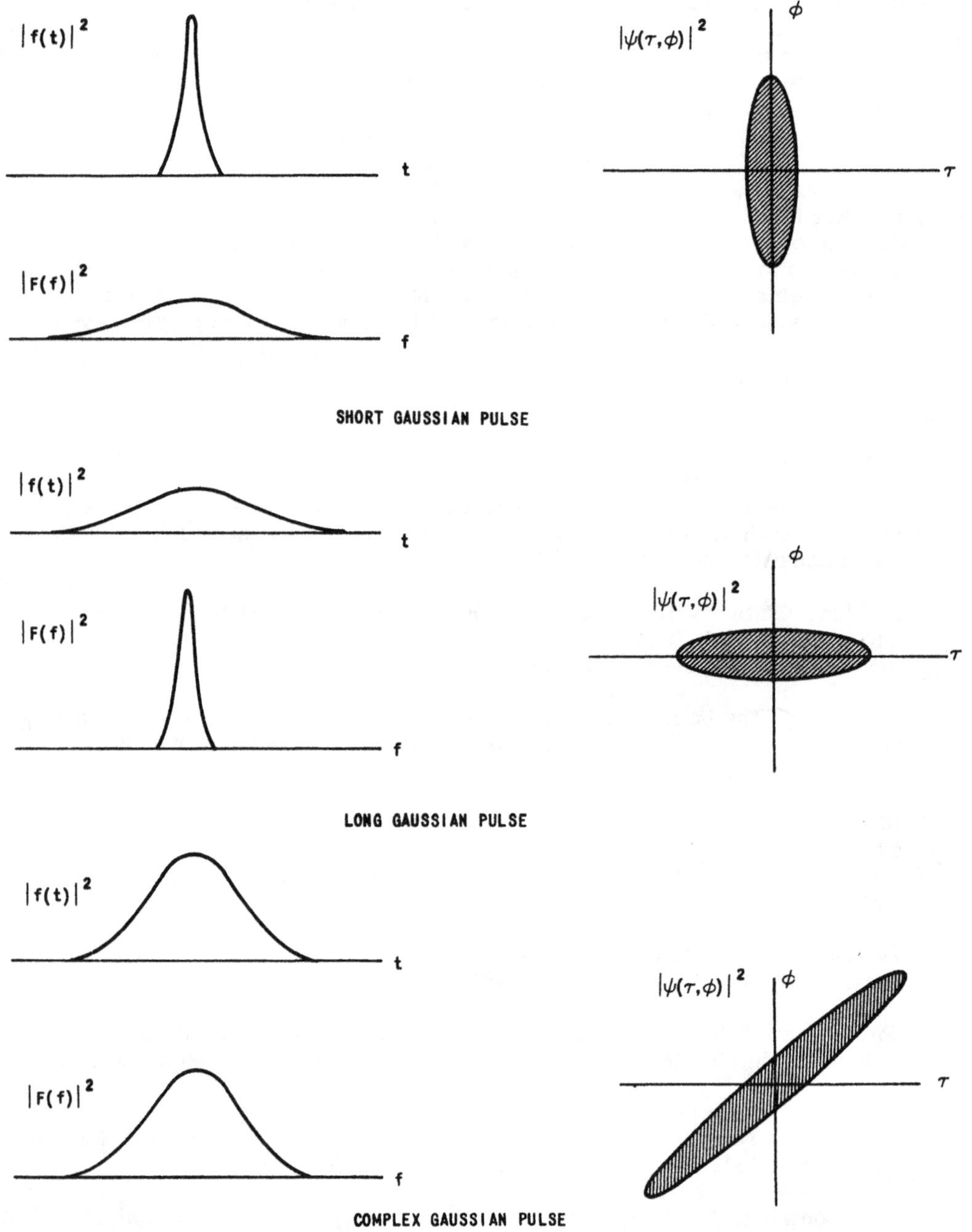

Figure A-7 - Joint correlation envelope for various Gaussian pulses

B. EFFECTS OF BOUNDS ON FUNCTIONS

1. INTRODUCTION

Thus far, we have discussed representing the structural properties of various functions without imposing restrictions upon them. However, restrictions are inherent characteristics of all real signals and a meaningful analysis is one which accounts for their effect on observed results. The fundamental restrictions are those imposed by finite bandwidth and finite time duration, either separately or together.

The imposition of a bound on a continuous function implies transforming it into a set of discrete data comprising a finite or an infinite number of coefficients. This is called sampling. For a one-to-one transformation to exist where each function will correspond to one and only one sequence of coefficients, the functions under consideration must be properly restricted. It is these restrictions that permit a function to be determined in any domain when only a related set of discrete data is known. A continuous interpolate may be constructed from the discrete data to give a satisfactory replica of the original function. The interpolation technique selected will depend on the properties of the actual function to be distinguished, and in turn on the criteria used, such as minimizing the mean-square error or minimizing the maximum error.

The primary purpose here is to show the basic similarities and differences of useful representations when the functions are bounded in time, frequency, or jointly. Unbounded functions contain a finite amount of power whereas bounded functions are "energy bounded." The selection of a particular description will depend not only on the specific application but also on the types of bounds imposed. Improved visualization, or easier computation and more economical instrumental realization can result from the proper selection of a description.

2. TIME LIMITS

UNIFORM SAMPLING

A function $f(t)$ which is continuous in a finite time interval and zero elsewhere can be represented by discrete values in terms of its spectrum. Owing to its importance in information processing, this relationship will be derived following the presentation employed by Woodward (Ref. B-2.1). The following steps are taken:

1. representing a function in a particular interval by a periodic function having the length of the interval as its period,

2. representing the Fourier spectrum of the original function by the frequency composition of the periodic function, and finally,

3. introducing an auxiliary function and developing a sampling theorem and associated composing function in the frequency domain.

Let $f_p(t)$ be a periodic function having a period equal to $f(t)$ in the interval $-T/2 \leq t \leq T/2$. Then

$$f_p(t) = \sum_{n=-\infty}^{\infty} D_n e^{j \frac{2\pi}{T} nt} \tag{A-6}$$

where

$$D_n = \frac{1}{T}\int_{-T/2}^{T/2} f_p(t)\, e^{-j\frac{2\pi}{T}nt} \tag{A-7}$$

Due to the above restriction $f(t)$ is completely determined by $f_p(t)$ and consequently its Fourier spectrum $F(j\omega)$ is given by the frequency composition of the periodic function:

$$F(j\omega) = F_p(j\omega) = \sum^{n} D_n\,. \tag{B-1}$$

Since the Fourier spectrum $F(j\omega)$ of $f(t)$ is formulated as

$$F(j\omega) = \int_{-\infty}^{\infty} f(t)\, e^{-j\omega t}\, dt \tag{A-11}$$

$$= \int_{-T/2}^{T/2} f_p(t)\, e^{-j\omega t}\, dt \tag{B-2}$$

by comparing it with (A-7) it is evident that D_n may be expressed in terms of $F(j\omega)$:

$$D_n = \frac{1}{T} F(j\omega)\Big|_{\omega = n\omega_o = \frac{2\pi n}{T}} \tag{B-3}$$

$$= \frac{1}{T} F\left(j\,\frac{2\pi n}{T}\right). \tag{B-4}$$

This shows that the spectrum of $f(t)$ is proportional to the coefficients D_n in the Fourier series representation of $f_p(t)$ for values $f = n/T$. The spectrum of $f_p(t)$ is a line spectrum:

$$F_p(j\omega) = \sum^{n} D_n = \frac{1}{T}\sum^{n} F\left(j\,\frac{2\pi n}{T}\right) = \frac{1}{T}\sum^{n} F(j\omega)\,\delta\left(\omega - \frac{2\pi n}{T}\right). \tag{B-5}$$

From (B-1) and B-5) it is seen that the spectrum of $f(t)$ may be expressed as a series of impulses at values of frequency equal to n/T whose strength is that of the spectrum evaluated at the corresponding frequency divided by T, where T is the duration of the waveform, $f(t)$.

By introducing another auxiliary function, $f(t)$ may be represented by $f_p(t)$ for all values of time, that is,

$$f(t) = f_p(t)\, g(t) \tag{B-6}$$

where $f_p(t)$ is now given by

$$f_p(t) = \frac{1}{T}\sum_{n=-\infty}^{\infty} F\left(j\,\frac{2\pi n}{T}\right) e^{j\frac{2\pi n}{T}t} \tag{B-7}$$

and $g(t)$ is defined as

$$g(t) \equiv \begin{cases} 1 & \text{for } |t| \le T/2 \\ 0 & \text{elsewhere} \end{cases} \tag{B-8}$$

whose spectrum $G(j\omega)$ is

$$G(j\omega) = \frac{\sin(\omega T/2)}{\omega/2} . \tag{B-9}$$

From Fourier theory, the spectrum of a product of two time functions is the convolution of the two separate spectra. Thus

$$F(j\omega) = \int_{-\infty}^{\infty} F_p(j\beta)\, G(j\omega - j\beta)\, d\beta. \tag{B-10}$$

Substituting (B-5) and (B-9) into (B-10)

$$F(j\omega) = \frac{1}{T}\int_{-\infty}^{\infty} \sum^{n} F(j\beta)\,\delta\left(\beta - \frac{2\pi n}{T}\right) \frac{\sin(\omega-\beta)\frac{T}{2}}{\frac{1}{2}(\omega-\beta)} d\beta \tag{B-11}$$

which results in

$$F(j\omega) = \sum^{n} F\left(j\,\frac{2\pi n}{T}\right) \frac{\sin\left(\omega - 2\pi\frac{n}{T}\right)\frac{T}{2}}{\frac{T}{2}\left(\omega - 2\pi\frac{n}{T}\right)} . \tag{B-12}$$

This is the sampling theorem in the frequency domain and expresses a continuous spectrum in terms of an equivalent line spectrum.

In general, if a time-limited signal is zero everywhere in the range $T_1 < t < T_2$, then its Fourier spectrum may be completely determined by giving its values at an infinite set of sample points spaced $1/(T_2 - T_1)$ cps apart in frequency. Sampling the spectrum of a time-limited waveform at this rate is equivalent to expressing it in terms of the coefficients of a Fourier series expansion of the waveform. The spacing between sample values controls the largest value of the conjugate variable (time), while the number of samples determines the order of the highest harmonic term in the Fourier series expansion.

INSTANTANEOUS POWER SPECTRUM

If a stationary random function is suddenly applied to a network and it is required to determine its power spectrum at some time T after it started, then if T is not sufficiently large the spectrum measured will not be identical with the power spectrum represented in section A.3. Instead, it will change with time until it is no longer dependent upon the starting time of the process. This behavior may be represented through the concept of the instantaneous power spectrum, as defined by Page (Ref. B-2.4).

Denote the energy density in the time-frequency plane by $W(t,f)$. Then the total energy expended up to a time T will be given by

$$\int_{0}^{T}\int_{-\infty}^{\infty} W(t,f)\, df\, dt . \tag{B-13}$$

The instantaneous power is defined as the rate of increase of the total energy and is thus expressed by differentiating (B-13),

$$\text{instantaneous power} = \int_{-\infty}^{\infty} W(T,f)\, df . \tag{B-14}$$

Comparing (B-14) with (A-48), $W(T, f)$ is defined as the instantaneous power spectrum at any instant T.

If a signal $f(t)$ is switched off at time $t = T$, in order to apply Fourier analysis define the "running transform" of $f(t)$ as the transform of a continually changing auxiliary function $f_t(x)$ such that:

$$f_t(x) \equiv \begin{cases} f(x) & \text{for} \quad x < t \\ 0 & \text{for} \quad x > t \end{cases} \tag{B-15}$$

$$S_t(f) = \int_{-\infty}^{\infty} f_t(x)\, e^{-j2\pi fx} dx = \int_{-\infty}^{t} f(x)\, e^{-j2\pi fx} dx \,. \tag{B-16}$$

The auxiliary signal will be identical with $f(t)$ up to time T, and therefore will deliver the same energy as $f(t)$. It should be noted that in (B-15) and (B-16), x is the variable of integration and t is any time T. As a result of Plancherel's energy theorem and (B-14), the instantaneous power spectrum must satisfy the following relationship

$$\int_{-\infty}^{t} W(x, f)\, dx = |S_t(f)|^2 \tag{B-17}$$

which is sufficient to determine $W(t, f)$. Differentiating (B-17) with respect to time gives:

$$W(t,f) = (\partial/\partial t)\,|S_t(f)|^2 \tag{B-18}$$

$$= 2f(t) \int_{-\infty}^{t} f(x) \cos 2\pi f(x - t)\, dx \,. \tag{B-19}$$

The mean power spectrum $\overline{W(f)}$ of a member $f(t)$ of an ergodic random process $[f(t)]$ may be expressed as

$$\overline{W(f)} = 2 \int_{0}^{\infty} \overline{f(t)\, f(t+\tau)} \cos 2\pi f\tau\, d\tau \,. \tag{B-20}$$

By placing $f(t)$ inside the integrand in (B-19), changing variables, and assuming $f(t)$ to be switched on at $t = 0$, (B-19) becomes

$$W(t,f) = 2 \int_{0}^{t} f(t)\, f(t+\tau) \cos 2\pi f\tau d\tau \,. \tag{B-21}$$

If (B-21) is averaged over the ensemble,

$$\overline{W(t,f)} = 2 \int_{0}^{t} \overline{f(t)\, f(t+\tau)} \cos 2\pi f\tau\, d\tau \,. \tag{B-22}$$

Since the time and statistical averages of an ergodic process are equivalent comparison of (B-22) with (B-20) shows that the stochastic average of the instantaneous power spectrum asymptotically approaches $\overline{W(f)}$. Equation (B-22) shows how the power spectrum of the process develops in time.

The instantaneous power spectrum $W(t, f)$ is not unique. Reference B-2.5 shows that it can have added to it a complementary function of frequency $W_c(t, f)$ satisfying

$$\int_{-\infty}^{\infty} W_c(t,f)\,df = 0 \tag{B-23}$$

without changing the original signal. If the instantaneous power spectrum of the same signal is derived by two independent observers, then at any time during the period of observation common to both observers the instantaneous power spectra may differ by the complementary function. Conversely, if over any interval of time, two instantaneous power spectra differ only by a complementary function, then the corresponding signals are identical during the interval of time.

RUNNING AUTOCORRELATION FUNCTION

It is now of interest to examine the relationship the instantaneous power spectrum has with a correlation function which denotes the time bound. Using (B-18) we get

$$W(t,f) = (\partial/\partial t)\,|S_t(f)|^2 \tag{B-18}$$

$$= (\partial/\partial t)\, S_t(f)\, S_t^*(f)\,. \tag{B-24}$$

From (B-16) we have,

$$W(t,f) = (\partial/\partial t) \int_{-\infty}^{\infty} f_t(x)\, e^{-j2\pi f x}\,dx \int_{-\infty}^{\infty} f_t(s)\, e^{+j2\pi f s}\,ds \tag{B-25}$$

$$= (\partial/\partial t) \int_{-\infty}^{\infty}\int_{-\infty}^{\infty} f_t(x)\, f_t(s)\, e^{-j2\pi f(x-s)}\,dx\,ds\,. \tag{B-26}$$

The auxiliary function $f_t(x)$ is used and not the actual signal $f(x)$ so the integrals above will have infinite limits. If we let $x-s = \tau$, (B-26) becomes

$$W(t,f) = (\partial/\partial t) \int_{-\infty}^{\infty}\int_{-\infty}^{\infty} f_t(x)\, f_t(x-\tau)\, e^{-j2\pi f\tau}\,dx\,d\tau \tag{B-27}$$

$$= (\partial/\partial t) \int_{-\infty}^{\infty} \left[\int_{-\infty}^{\infty} f_t(x)\, f_t(x-\tau)\,dx\right] e^{-j2\pi f\tau}\,d\tau\,. \tag{B-28}$$

With reference to (A-47), the bracketed term in (B-28) is the temporal (finite energy) autocorrelation function of $f_t(x)$ or the "running autocorrelation function" $\psi_t(\tau)$ of the signal $f(x)$. Placing $(\partial/\partial t)$ inside the integrand, $W(t,f)$ may now be expressed as

$$W(t,f) = \int_{-\infty}^{\infty} \left[(\partial/\partial t)\,\psi_t(\tau)\right] e^{-j2\pi f\tau}\,d\tau \tag{B-29}$$

and is the Fourier transform of the time rate of change of the running autocorrelation function. This is the Wiener-Khintchine relationship as applied to energy-bounded functions having time-dependent power spectra.

If the signal is a nonstationary time series, ensemble rather than time averaging must be employed. The bracketed term in (B-28) can no longer represent the autocorrelation function. If we take the statistical average of both sides of (B-27) and carry out the indicated differentiation, we have

$$\overline{W(t,f)} = \int_{-\infty}^{\infty} \psi(t,\tau)\, e^{-j2\pi f\tau}\, d\tau \tag{B-30}$$

where

$$\psi(t,\tau) = \overline{f(t)\; f(t-\tau)}. \tag{B-31}$$

Equation (B-30) is the Wiener-Khintchine theorem as applied to power-bounded nonstationary waveforms.

ORTHOGONALIZED EXPONENTIALS

A sinusoidal representation describes signal ensembles whose amplitudes do not vary with time. Moreover, a decomposition into sinusoidal components is appropriate whenever it is desired for the elementary signals to be invariant under translation in time, that is, when we do not have to characterize the epochs or instants of time at which the signal components are created. A Fourier representation is very useful in describing the time averaged properties of a system or signal ensemble. However, it is not well suited for discriminating between signals or for detecting the occurrence of a particular signal against the background of a noise ensemble.

It is often necessary to characterize a signal by both its epoch and structure. Specifying the epoch is required in determining phase information. Analysis of information-bearing signals is also concerned with a discrete representation of low dimensionality and a means of evaluation of performance which relates physical measurements with mathematical theory. A method for obtaining these features is to represent a signal by specifying the generator of that signal. Since this entails an understanding of the basic physical mechanisms involved in the generation of the signal, the parameters of the representation will acquire a deeper significance and meaning.

If $f(t)$ is an impulsive excitation to a signal generator characterized by the response $h(t,\tau)$ to a unit impulse applied at time τ, the resulting response $s(t)$, or desired signal, may be expressed as

$$s(t) = \int_{0}^{\infty} f(\tau)\, h(t;\tau)\, d\tau\ . \tag{B-32}$$

The problem in signal analysis would be to recover from the observed signal $s(t)$ a specification of the excitation function $f(\tau)$ and of the system function $h(t,\tau)$.

The signal generator impulse response $h(t,\tau)$ is best characterized by its natural frequencies p_k which appear in the exponent of terms having the form

$$A_k\, e^{p_k(t-\tau)} \qquad (t > \tau) \tag{B-33}$$

where τ is the epoch. In practice, the p_k parameters are quite difficult to determine, whereas the amplitude coefficients are easily determined once the exponential factors are known. This suggests selecting another set of exponentials s_k to approximate the original set, leaving the epochs and amplitude coefficients to specify the impulse response. For a class of functions formed from damped exponential components, such as (B-33), a small number of preselected damped exponentials which cover a region in the left side of complex-frequency plane may provide a good approximation to any exponential having a frequency within that region. No such approximation is possible for a class of functions formed from sine waves.

The problem of characterizing the system function $h(t,\tau)$ is to establish an appropriate, discrete set of exponentials that will approximate with allowable error over a semi-infinite time interval the actual system function, i.e.,

$$h(t,\tau) = \sum_k A_k(\tau)\, e^{s_k(t-\tau)} \qquad (t \geq \tau)$$

$$= 0 \qquad (t < \tau) \,. \qquad \text{(B-34)}$$

The variation of $h(t,\tau)$ is then accounted for by the variation of the amplitude coefficients $A_k(\tau)$. The error criterion used is the mean-square error. It is desired that the component functions be uncorrelated or orthogonalized, otherwise a change in the amplitude of one may be more or less neutralized by changes in the amplitudes of other components.

In general, a frequency-domain analysis implies the specification of the amplitudes of many different waveshapes, all having the same epoch. A time domain analysis, however, specifies the amplitudes of many different components, all having the same waveshape but differing in time of occurrence. It is readily seen that a representation using orthogonalized exponentials is a time-domain representation.

To conclude, the method discussed represents a signal by the convolution of two functions. The first characterizes the temporal attributes of the signal by the epochs and intensities of the impulses comprising the function. The second function characterizes the structural attributes of the signal by the impulse response of a generator. The impulse response is then approximated by a sum of a number of orthogonalized exponentials corresponding to the natural modes of the generator.

REFERENCES

B-2.1. P. M. Woodward, "A Mathematical Description of Random Noise," I.R.E. Journal, October 1948
(Discusses representing a continuous spectrum by a line spectrum)

B-2.2. S. Goldman, "Information Theory," Chapter 2, Prentice-Hall, Inc., 1953
(Discusses sampling in the frequency domain)

B-2.3. W. B. Davenport, Jr. and W. L. Root, "An Introduction to the Theory of Random Signals and Noise," Chapter V, McGraw-Hill Company, Inc., 1958

B-2.4. C. H. Page, "Instantaneous Power Spectra," Journal Appl. Phys., Vol. 23, No. 1, pp. 103-106, January 1952

B-2.5. C. H. M. Turner, "On the Concept of an Instantaneous Power Spectrum, and its Relationship to the Autocorrelation Function," Journ. Appl. Phys., Vol. 25, No. 11, pp. 1347-1351, November 1954

B-2.6. D. G. Lampard, "Generalization of the Wiener-Khintchine Theorem to Non-stationary Processes," Journ. Appl. Phys., Vol. 25, No. 6, June 1954

B-2.7. W. H. Huggins, Part 1, "The Use of Orthogonalized Exponentials," ASTIA Document No. AD133741, September 30, 1957

B-2.8. W. H. Kautz, "Transient Synthesis in the Time Domain," I.R.E. Trans., CT-1, No. 3, pp. 29-39, September 1954

3. FREQUENCY LIMITS

BANDLIMITED FOURIER SERIES

Fourier series representation converts a continuous function having a finite number of discontinuities into an infinite series of discrete terms. When the series is bandlimited it comprises a finite number of terms such that the coefficients A_n and B_n are zero for indices greater than a given number N. It is necessary only to consider the first $2N + 1$ coefficients. For period T, frequency $(N+1)/T$ has a zero coefficient and N/T a coefficient of finite value. The limit in frequency W is set halfway between the two:

$$W = \left(\frac{N+1}{T} + \frac{N}{T}\right)\frac{1}{2} \tag{B-35}$$

$$= \frac{N + 0.5}{T} \text{ cps.} \tag{B-36}$$

The bandlimited Fourier series can then be considered as a finite vector representation containing $2N + 1$ or $2WT$ coefficients.

A waveform periodic with period T, when limited to a band of frequencies less than W, may be represented by $2WT$ coefficients of its Fourier series. Clearly, $2WT$ values, properly restricted, of the waveform itself will also be a representation. There are a number of sets of $2WT$ values that form representations which will allow the whole waveform $f(t)$ to be recovered for the period T. The simplest of the reconstruction series is the one having a "sine-over-sine" composing function, i.e.,

$$f(t) = \sum_{n=1}^{2WT} f\left(\frac{n}{2W} - t_o\right)\left[\frac{\sin \pi\left[2W(t+t_o) - n\right]}{2WT \sin \dfrac{\pi\left[2W(t+t_o) - n\right]}{2WT}}\right] \tag{B-37}$$

where the coefficients of the composing functions are taken at time intervals spaced $1/2W$ seconds apart starting at some time less than $1/2W$, $t_o (0 \le t_o \le 1/2W)$ (Figure B-1(a)).

UNIFORM SAMPLING

If the waveform under consideration has a Fourier transform it can be represented by orthogonal "sin x-over-x" sampling functions. This is Shannon's sampling theorem (Ref. B-3.2), and states that a bandlimited signal whose Fourier spectrum contains no component above frequency W cps, is uniquely determined by its values at $1/2W$-seconds apart. By applying the sampling theorem to a signal $f(t)$ whose Fourier spectrum is $F(j\omega)$, the reconstruction of the signal is given by:

$$f(t) = \sum_{n=-\infty}^{\infty} f\left(\frac{n}{2W}\right)\phi_n(t). \tag{B-38}$$

The function $\phi_n(t)$ is called the composing function for the sample point $t = n/2W$ and is given as:

$$\phi_n(t) = \frac{\sin \pi(2Wt - n)}{\pi(2Wt - n)} \tag{B-39}$$

The Fourier spectrum of the signal can be expressed in terms of the uniform sample values according to

SAMPLING PROCESS	SAMPLING THEOREM
(a) FOURIER SERIES BAND-LIMITED	If a band-limited Fourier series containing 2WT coefficients can represent a waveform, 2WT values, properly restricted, of the waveform itself will also be a representation. The simplest of the many possible reconstruction series is the one having a "sine-over-sine" com ing function where the coefficients are taken at time intervals spaced 1/2W seconds apart st ing at some time less than 1/2W, t_o.
(b) UNIFORM: LOW-PASS	If the waveform under consideration has a Fourier transform it can be represented by an orthogonal set of "sin x over x" sampling functions. Shannon's sampling theorem states that a band-limited (0,W) signal whose Fourier spectrum contains no component above frequency W cycles per second is uniquely determined by its values at an infinite set of sample points spaced at 1/2W seconds apart.
(c) UNIFORM: HIGH-PASS	If a complex waveform f(t) has its Fourier spectrum confined within the positive freque interval $\left(f_o - \frac{1}{2}W,\ f_o + \frac{1}{2}W\right)$ where $f_o \geq \frac{1}{2}W$, then it can be uniquely determined by its comp values at intervals 1/W. If f(t) = g(t) + jh(t), then the purely real waveform g(t) may be recovered by specifying the amplitude $\sqrt{g(t)^2 + h(t)^2}$ and the instantaneous phase angle tan at each sampling point; i.e., the envelope and phase of the carrier.
(d) DERIVATIVE	If a function f(t) contains no frequency higher than W cycles per second, it may be determined by evaluating the function amplitude and derivatives at an infinite set of sampl points spaced (K+1)/2W seconds apart, where K is the order of the highest derivative when all lower ordered derivatives are observed in each sample.

Figure B-1 - Several periodic sampling techniques for bandlimited waveforms

RECONSTRUCTION FUNCTION $f(t) = \Sigma_n f(t_n)\, \phi_n(t)$	COMPOSING FUNCTION $\phi_n(t)\big\vert_{n=o}$	FOURIER TRANSFORM OF COMPOSING FUNCTION
) $= \sum_{n=1}^{2WT} f\left(\frac{n}{2W} - t_o\right) \left[\frac{\sin \pi \left[2W(t+t_o) - n\right]}{2WT\, \frac{\sin \pi \left[2W(t+t_o) - n\right]}{2WT}}\right]$ $o \leq t_o \leq \frac{1}{2W}$	$\frac{\sin 2\pi Wt}{2WT \sin \frac{\pi t}{T}}$ 1 t $t_o = 0$ 0	$F\left\{\frac{\sin 2\pi Wt}{2WT \sin \frac{\pi t}{T}}\right\}$ ω 0
t) $= \sum_{n=-\infty}^{\infty} f\left(\frac{n}{2W}\right) \frac{\sin \pi\,(2Wt - n)}{\pi\,(2Wt - n)}$	$\frac{\sin 2\pi Wt}{2\pi Wt}$ 1 t 0 $\frac{1}{2W}$ $\frac{1}{W}$ $\frac{3}{2W}$	$F\left\{\frac{\sin 2\pi Wt}{2\pi Wt}\right\}$ 1/2 W $-2\pi W$ $2\pi W$ ω 0
$= \sum_{n=-\infty}^{\infty} f\left(\frac{n}{W}\right) \frac{\sin \pi\,(Wt - n)}{\pi\,(Wt - n)} \exp\left\{j2\pi f_o\left(t - \frac{n}{W}\right)\right\}$	$\frac{\sin \pi Wt}{\pi Wt}$ exp. $j\omega_o t$ 1 t slope $= \omega_o$ 0 $\frac{1}{W}$ $\frac{2}{W}$ $\frac{3}{W}$	$F\left\{\frac{\sin \pi Wt}{\pi Wt} \exp j\omega_o t\right\}$ $\frac{1}{W}$ 0 $2\pi f_o$ ω $2\pi\left(f_o - \frac{W}{2}\right)$ $2\pi\left(f_o + \frac{W}{2}\right)$
t derivative: K = 1 $= \sum_{n=-\infty}^{\infty} \left[f\left(\frac{n}{W}\right) + \left(t - \frac{n}{W}\right) f'\left(\frac{n}{W}\right)\right] \left[\frac{\sin \pi\,(Wt - n)}{\pi\,(Wt - n)}\right]^2$	K = 1 $\left(\frac{\sin \pi Wt}{\pi Wt}\right)^2$ 1 t $\frac{1}{W}$ $\frac{2}{W}$ $\frac{3}{W}$ $\frac{4}{W}$ 0	K = 1 $F\left\{\left(\frac{\sin \pi Wt}{\pi Wt}\right)^2\right\}$ $\frac{1}{W}$ $-2\pi W$ $2\pi W$ ω 0

$$F(j\omega) = \frac{1}{2W}\sum_{n=-\infty}^{\infty} f\left(\frac{n}{2W}\right) e^{-jn\omega/2W} \qquad |\omega| \le 2\pi W$$

$$= 0 \qquad |\omega| > 2\pi W . \tag{B-40}$$

It is seen that sampling a bandlimited function at the Shannon rate is equivalent to expressing it in terms of the coefficients of a Fourier series expansion of the spectrum (Figure B-1(b)).

The process of periodically sampling a function $f(t)$ instantaneously is the same as multiplying the function by a train of impulses of unit area which are spaced uniformly at intervals equal to the sampling time, i.e.,

$$f(t)^* = \sum^{n} f(t)\,\delta(t - n/2W) \tag{B-41}$$

where $f(t)^*$ is the sampled series. The waveform can finally be recovered passing $f(t)^*$ through a low-pass filter whose impulse response $h(t)$ is

$$h(t) = \frac{\sin 2\pi Wt}{2\pi Wt} = \phi_n(t)\Big|_{n=0} \tag{B-42}$$

A complex waveform $f(t)$ whose spectrum is confined within the positive frequency interval $(f_o - 1/2W,\ f_o + 1/2W)$ may be uniquely determined by its complex values at intervals $1/W$. The reconstruction formula for this case is given by

$$f(t) = \sum_{n=-\infty}^{\infty} f\left(\frac{n}{W}\right)\frac{\sin \pi(Wt - n)}{\pi(Wt - n)} \exp\left\{j2\pi f_o\left(t - \frac{n}{W}\right)\right\}. \tag{B-43}$$

In order to apply (B-43) to a real waveform, it is only necessary to take the real part of both sides. If $f(t) \equiv g(t) + jh(t)$ we obtain for $g(t)$

$$g(t) = \sum_{n=-\infty}^{\infty} g\left(\frac{n}{W}\right)\frac{\sin \pi(Wt - n)}{\pi(Wt - n)} \cos 2\pi f_o(t - n/W)$$

$$- \sum_{n=-\infty}^{\infty} h\left(\frac{n}{W}\right)\frac{\sin \pi(Wt - n)}{\pi(Wt - n)} \sin 2\pi f_o(t - n/W) . \tag{B-44}$$

Equation (B-44) means that an amplitude $\sqrt{g^2(t) + h^2(t)}$ and an instantaneous phase angle $\tan^{-1}(h/g)$ must be specified at each sampling point. They represent the amplitude and phase of the carrier. For a bandwidth W, $g(t)$ and $h(t)$ must both be specified at intervals $1/W$, and the total number of degrees-of-freedom in a high-frequency waveform in time T is $2WT$. This is the same as for a low-frequency waveform, although the sampling interval is twice the latter (Figure B-1(c)).

The concept of sampling also provides physical meaning to an exact bandwidth limitation. The Shannon expansion shows that a bandwidth-limited signal $f(t)$ is entirely defined by the sequence of values $f_1, f_2, \ldots, f_n, \ldots$ taken at regularly-spaced times $t_o,\ t_o + T,\ \ldots,\ t_o + nT$ (with $T = 1/2W$). The imposition of an exact bandwidth limit on a general signal implies that the only values of the signal taken into account are those at the uniformly-spaced times determined by the Shannon rate. This means that if different signals take on the same values at these times, they are no longer distinguishable after being filtered by an ideal filter. Mathematically, a bandwidth limitation permits a continuous function be replaced by an enumerable sequence. Physically it permits transmission of information more economically with a conservation of bandwidth.

Two sampling theorems that find extreme importance in signal processing when the samples are taken at regular intervals and are independent of the exact times of sampling are the following:

1. Encipherment

The magnitude of each sample may be varied in an arbitrary manner without increasing the frequency range of the samples.

2. Relation between mean square of signal and its samples

If the square of the signal does not contain a component of frequency equal to some integral multiple of the sampling frequency or two components such that their absolute sum or difference are also some integral multiple of the sampling frequency, then the sum of the average of the squares of the samples will be equal to the mean-square value of the sampled wave.

REFERENCES

B-3.1. T. G. Birdsall, "Sampling Theorems," Lecture Notes (unpublished) (Discusses bandlimited Fourier series)

B-3.2. S. Goldman, "Information Theory," Prentice Hall, Inc., 1953 (Chapter 2 discusses time and frequency sampling of continuous waveforms and Chapter 8 discusses constraints on the amplitude)

B-3.3. D. L. Jagerman and L. F. Fogel, "Some General Aspects of the Sampling Theorem," I.R.E. Trans., Vol.IT-2, pp. 139-145, December 1956

B-3.4. A. V. Balakrishman, "A Note on the Sampling Principle for Continuous Signals," I.R.E. Trans., Vol.IT-3, pp. 143-146, June 1957

B-3.5. H. S. Black, "Modulation Theory," Chapter IV, D. Van Nostrand Company, Inc., 1958 (Discussion of sampling principles)

B-3.6. D. W. Tufts, "A Sampling Theorem for Stationary Random Process," Quarterly Progress Report, pp. 87-93, Research Laboratory of Electronics, M.I.T., April 15, 1959

B-3.7. P. M. Woodward, "Probability and Information Theory, with Applications to Radar," pp. 34-35, Pergamon Press, 1957

4. JOINT TIME AND FREQUENCY LIMITS

INTRODUCTION

Thus far, we have discussed descriptions of functions bounded in time or frequency and functions not restricted in either of the domains. The descriptions were found to depend on the type of functions considered. For an unbounded function, periodic or random, the energy content is infinite which suggests that a power description is more meaningful. A transient contains a finite amount of energy which implies that we use energy descriptions. When the functions are time or frequency limited, the concept of sampling permits obtaining a complete representation. A time-limited function would be represented completely by an infinite number of samples of the spectrum. Similarly, a frequency-limited function would theoretically extend over all time and would be represented by an infinite number of time samples. When the function is both time and bandlimited, we can no longer represent the function by an infinite number of independent samples in either domain. The representation will always be an approximation, its value depending on the number and type of elementary functions used in the decomposition. Therefore, the primary difference between a function time or bandlimited and one limited both in time and frequency is that the former may be represented either approximately or completely, while the latter can only be represented through an approximation.

Theoretically, it is not possible to construct a Fourier pair, not equivalent to zero, which has the property that the function and its transform both vanish in any finite intervals of the conjugate variables. This will be shown following the proof by Wernikoff (Ref. B-4.1).

If a signal is bandlimited such that its spectrum $F(f)$ vanishes outside the frequency interval $(-W, W)$, i.e.,

$$F(f) = 0 \qquad \text{for} \quad f > |W| \tag{B-45}$$

it can be represented by an infinite sum of $\sin x/x$ functions:

$$f(t) = \sum_{n=-\infty}^{\infty} a_n \frac{\sin \pi(2Wt - n)}{\pi(2Wt - n)} \tag{B-46}$$

where a_n is determined by examining the signals at times, t_n, separated by intervals of $1/2W$. That is

$$a_n = f(t_n) = f\left(\frac{n}{2W}\right) . \tag{B-47}$$

If $f(t)$ is limited to $(0,T)$, the only samples that are not zero are those taken in $(0,T)$. Then $f(t)$ will be given by a finite sum of $\sin x/x$ functions, so that (B-46) becomes

$$f(t) = \sum_{n=0}^{N} a_n \frac{\sin \pi(2Wt - n)}{\pi(2Wt - n)} \tag{B-48}$$

where $N = 2WT$ samples. However, the assumption that $f(t)$ is limited to $(0,T)$ implies that

$$\sum_{n=0}^{N} a_n \frac{\sin \pi(2Wt - n)}{\pi(2Wt - n)} = 0 \tag{B-49}$$

for all t outside the interval $(0,T)$. This requirement states that the tails of a finite number of $\sin x/x$ functions have to combine in such a way that they cancel each other completely outside the interval. Since the $\sin x/x$ functions are linearly independent over any interval, there cannot exist a set of nonvanishing coefficients a_n that satisfy this requirement. That is, there does not exist any function, not equivalent to zero, limited simultaneously in time and in frequency.

Representations of signals which last for time T and occupy bandwidth W may however be approximated by $2WT$ coefficients. Since the $\sin x/x$ function falls off slowly, the sample points determined outside the interval T may affect the signal inside the interval T. It was indicated that it is not possible to describe exactly a function which has a finite time duration and finite spectral bandwidth. Actually, physical processes are characterized by having most of the energy confined within a finite time duration and finite bandwidth. If T is the approximate duration of the signal and W is its approximate spectral bandwidth, the signal may be reconstructed to a high degree of accuracy by its values at $2WT$ sampling points provided that $2WT \gg 1$. This product represents the same number of values needed to recover a function that can be represented by a bandlimited Fourier series.

Sampling is an interpolation and the sampled representation of the function improves with increase in the number of sample data. When considering only a finite number of samples, due to sampling over a finite interval of time T, there will be an interpolation error which may become quite large at the edges of the interval. The error inside the interval will be reduced as the 2WT product is increased. 2WT samples reconstruct a waveform uniquely only if the waveform is a repetitive function of period T and the samples are all taken during one period. Nonperiodic waveforms analyzed for a time interval equal to T will have energy associated with the function beyond this interval. The magnitude of this energy is a measure of the interpolation error.

ELEMENTARY SIGNALS

A time and bandlimited function may be decomposed into a finite sum of elementary signals. An elementary signal $\phi(t)$, may have no restrictions other than that of having finite energy and preferably a low TW product. Consider a collection of elementary signals $\phi_n(t)$ defined as the successive time translates of $\phi(t)$ by a time interval θ:

$$\phi_n(t) = \phi(t - n\theta) \qquad \text{(B-50)}$$

where n is an integer which refers to the time instant t_n. If these signals are multiplied by the periodic exponential factors $e^{j2\pi(m/\theta)t}$, a set of frequency and time translates of $\phi(t)$, $\phi_{mn}(t)$, will be generated. These functions are expressed as

$$\phi_{mn}(t) = \phi(t - n\theta)e^{j2\pi\frac{m}{\theta}t} . \qquad \text{(B-51)}$$

The effect of a unit change in index n is to shift the spectrum of $f(t)$ by a unit of $1/\theta$.

This set of functions is complete in the sense that almost any signal $f(t)$ may be written as

$$f(t) = \sum_{m,n} a_{mn}\phi_{mn}(t) . \qquad \text{(B-52)}$$

In a time-frequency plane, one possible representation is a checkerboard pattern whose lines cross the time-axis at intervals equal to θ, and the frequency-axis at intervals of $1/\theta$. One of the a_{mn} is associated with each rectangle or "cell" of the resulting checkerboard (Figure B-2(a)). The basic pattern of this representation is set not by the choice of $\phi(t)$, but by the choice of the translation interval (θ). Each θ will lead to a different way of breaking up a region of time and frequency. In Figure B-2(b), a TW region of area 9 units is divided equally in time and frequency. We can also choose θ so as to divide only the time interval as in Figure B-2(c). This corresponds to specifying a signal by its samples in the time domain. We might also have chosen θ to produce the representation shown by Figure B-2(d), that is, a representation in terms of frequency.

The values of the a_{mn} depend on the choice of $\phi(t)$ and of θ. However, if the duration of $\phi(t)$ is almost equal to θ and its bandwidth almost equal to $1/\theta$, then for any given choice of θ, the values of the a_{mn} do not vary significantly with the choice of $\phi(t)$. Consequently, the elementary function $\phi(t)$ is somewhat arbitrary. $\phi(t)$ may be selected for its analytic properties, or the response it produces in a given network, or the understanding it produces concerning a physical problem.

ANALYTIC SIGNALS

A particular elementary signal which leads to useful concepts, such as instantaneous frequency and instantaneous phase, is the "analytic signal." This is obtained by adding a signal, properly restricted, in quadrature to a real signal. A real signal $f(t)$ is one whose Fourier spectrum is completely determined if it is given only for positive frequencies, i.e.,

$$F(j\omega) = F^*(j\omega) \qquad \text{(B-53)}$$

where the asterisk denotes the complex conjugate, or equivalently

$$|F(j\omega)| = |F(-j\omega)|$$

$$\arg F(j\omega) = -\arg F(-j\omega) . \tag{B-54}$$

If a real function of time $x(t)$ has finite energy, that is,

$$\int_{-\infty}^{\infty} x^2(t)\,dt \quad \text{exists and is finite} \tag{B-55}$$

then we may associate a real function $y(t)$ with $x(t)$ such that the function $z(u) = x(u) + jy(u)$ is an analytic function of the variable $u = t + j\theta$. The major requirements for this pair of functions which make $z(u)$ analytic are:

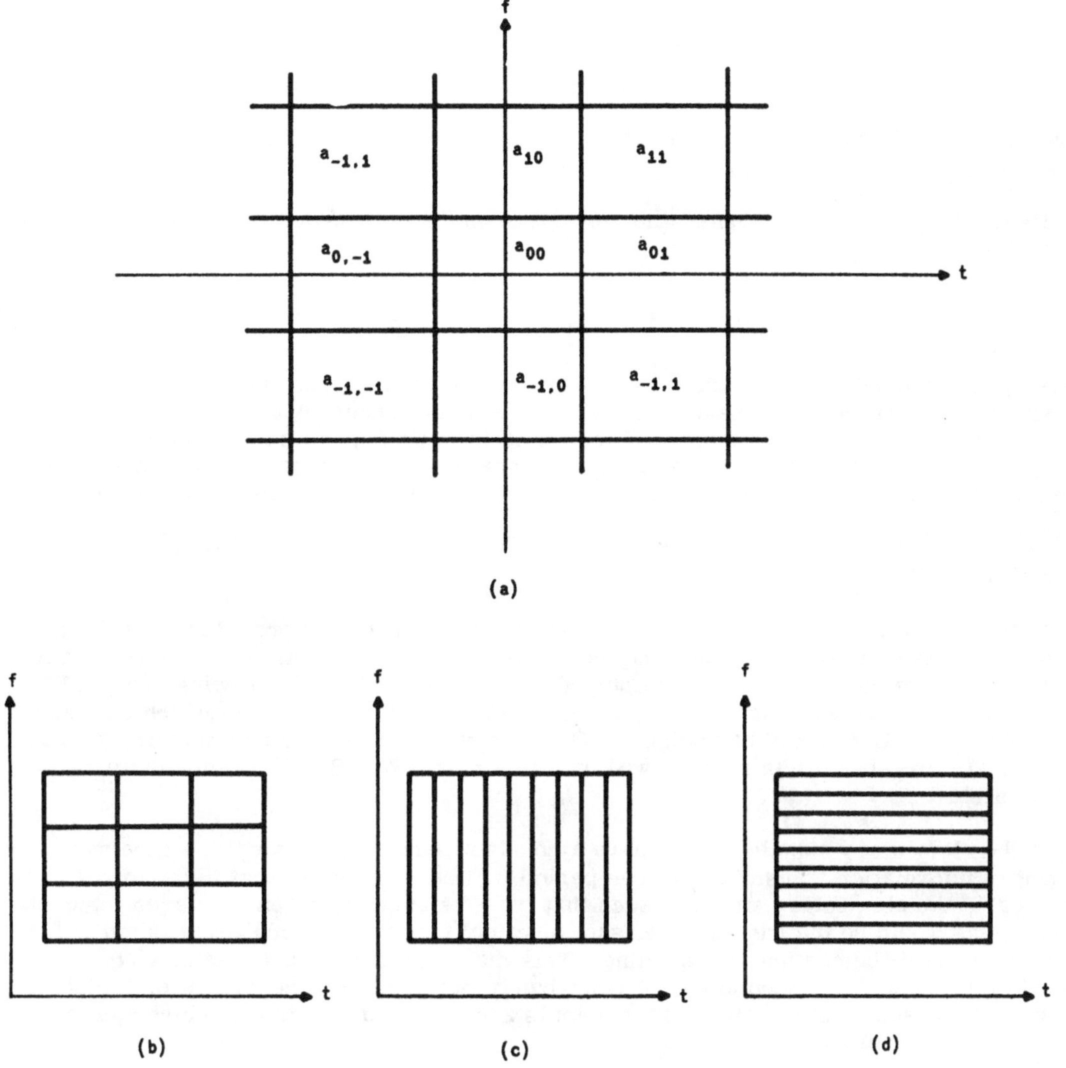

Figure B-2 - Time-frequency diagram of an arbitrary function

1. Each function of a pair has the same energy

$$\int_{-\infty}^{\infty} x^2(t)\,dt = \int_{-\infty}^{\infty} y^2(t)\,dt \,. \tag{B-56}$$

2. The functions are orthogonal

$$\int_{-\infty}^{\infty} x(t)\,y(t)dt = 0 \,. \tag{B-57}$$

3. The Fourier spectra $X(j\omega)$ and $Y(j\omega)$ of $x(t)$ and $y(t)$, respectively, are related by

$$\begin{aligned} Y(j\omega) &= -jX(j\omega) \qquad \text{when} \quad \omega > 0 \\ Y(j\omega) &= jX(j\omega) \qquad \text{when} \quad \omega < 0. \end{aligned} \tag{B-58}$$

Due to restriction 3, the Fourier spectrum $Z(j\omega)$ of $z(t) = x(t) + jy(t)$ has the properties

$$\begin{aligned} Z(j\omega) &= 2X(j\omega) \qquad \text{for} \quad \omega > 0 \\ &= 0 \qquad \text{for} \quad \omega < 0. \end{aligned} \tag{B-59}$$

Consequently, it is necessary to consider only the positive-frequency half of the Fourier frequency axis.

By defining the analytic signal, the concept of an "instantaneous frequency," $p(t)$, may be generalized, that is

$$p(t) \equiv \frac{1}{2\pi}\frac{d}{dt}\left[\arg z(t)\right] = \frac{1}{2\pi}\frac{dq(t)}{dt} \tag{B-60}$$

where $q(t) = \tan^{-1} y(t)/x(t)$ and may be regarded to be the "instantaneous phase." If $x(t)$ is a cosine wave, then $y(t)$ will be a sine wave of the same frequency and $f(t)$ will be constant, equal to the frequency of the signals. The definition in (B-60) is similar to that used in the theory of frequency modulation. The analytic signal may also be used to help distinguish the relationships between the effective time duration and bandwidth occupancy of signals.

TIME-BANDWIDTH PRODUCTS

The usefulness of the concept of time-bandwidth product is in providing an estimate of the number of degrees-of-freedom that may be required to specify a signal. The exact value will depend on the definition of duration employed and the particular signal being analyzed. It is often desired to resolve a signal into a series of elementary functions to which one, and only one, of these numbers could be assigned. One way of selecting these elementary waveforms is so that their time-bandwidth product will be a minimum (Ref. B-4.2). Other criteria may also be established.

If the elementary signals chosen have a constant period, for example, a sine wave, then the amount of information transmitted in one period will be identical to that in any other period. However, if the elementary signal is such that the "distance" between the zeros is constantly changing (as would be the case for a Bessel function) then the information transmitted during each interval will be continually changing. This may result in a whole "family" of time-bandwidth products. Consequently, the time-bandwidth product is dependent on the structure of the zero-crossings of the elementary signals and time-variant or time-invariant elements are needed for transmission.

Lampard (Ref. B-4.5), has shown that by expressing an "equivalent duration" $\Delta\tau$ and an "equivalent bandwidth" Δf in terms of the autocorrelation function $\psi(\tau)$ and power spectrum $W(f)$ of the signal, the identity

$$\Delta\tau \, \Delta f = 1 \tag{B-61}$$

is valid, provided the power spectrum of the signal extends down to zero frequency. For a transient signal and stationary time series:

$$\Delta\tau = \frac{\frac{1}{2}\int_{-\infty}^{\infty} \psi(\tau)\, d\tau}{\psi(0)} \tag{B-62}$$

and

$$\Delta f = \frac{2\int_{-\infty}^{\infty} W(f)\, df}{W(0)} . \tag{B-63}$$

The widths $2\Delta\tau$ and $1/2\Delta f$ represent the widths of the rectangles having the same areas as those under the correlation function and power spectrum, and having the same ordinate at $\tau = 0$ and $f = 0$, respectively. These definitions, (B-62), and (B-63), are particularly appropriate for cases in which $W(f) < W(0)$, for all f.

For the case of two nonstationary time series, we may use definitions of time dependent correlation functions and power spectra given in section B-2. The identity (B-61) is still valid where

$$\Delta\tau = \frac{\int_{0}^{\infty} \left[\psi_{12}(t,\tau) + \psi_{21}(t,\tau)\right] d\tau}{\left[\psi_{12}(t,0) + \psi_{21}(t,0)\right]} \tag{B-64}$$

and

$$\Delta f = \frac{\int_{0}^{\infty} \left[W_{12}(t,f) + W_{21}(t,f)\right] df}{\left[W_{12}(t,0) + W_{21}(t,0)\right]} . \tag{B-65}$$

It is interesting to note that though (B-64) and (B-65) are continuously changing, individually, for each instant of time, their product remains constant or time invariant. Thus, the time-bandwidth product presented here for a nonstationary time series is related to that when a signal is resolved into nonharmonic elementary signals. Equations (B-64) and (B-65) reduce to (B-62) and (B-63) if the time series approaches its steady-state behavior and both time series are equal.

We can also use a definition of duration analogous to that in quantum mechanics, namely

$$(\Delta x)^2 = \frac{\int_{-\infty}^{\infty} (x - x_o)^2 \, |s(x)|^2 \, dx}{\int_{-\infty}^{\infty} |S(x)|^2 \, dx} \tag{B-66}$$

where

$$x_o = \int_{-\infty}^{\infty} x|s(x)|^2 dx . \tag{B-67}$$

The x is an arbitrary variable, can be time, frequency, displacement, etc., Δx signifies a duration in x, and $s(x)$ is the Fourier description in the corresponding domain. In (B-66), Δx has the form of a standard deviation which in engineering terminology is an rms value. Equivalently, (B-66) may be considered as expressing the spread in x as the variance of $|s(x)|^2$; x_o would then represent a mean value.

If (B-66) is used to calculate the duration Δt and the corresponding radian bandwidth $\Delta\omega$ of a signal, where $s(x)$ is now the waveform and Fourier spectrum respectively, the Schwarz inequality may be used to give the result

$$\Delta t \cdot \Delta\omega \geq \frac{1}{2} . \tag{B-68}$$

Gabor (Ref. B-4.2), showed that the equality holds when the pulses are of Gaussian form; $\Delta\omega$ is then the radian bandwidth required to transmit a complete pulse in time Δt. If the functions $f(t)$ are real, then provided that $F(0) = 0$, (B-68) becomes

$$\Delta t \, \Delta\omega_+ > \frac{1}{2} \tag{B-69}$$

where $\Delta\omega_+$ is the variance of $|F_+(j\omega)|^2$, the square of the magnitude of the positive-frequency spectrum of $f(t)$. By replacing $f(t)$ by an analytic signal, $F_+(j\omega)$ is defined as

$$\begin{aligned} F_+(j\omega) &= 2F(j\omega) \qquad \omega \geq 0 \\ &= 0 \qquad\qquad \omega < 0 . \end{aligned} \tag{B-70}$$

Note the absence of the equality sign in (B-69). It was indicated above that an equality can be achieved for Gaussian pulses, but these have negative frequencies in their spectra which contradicts the assumption of an analytic signal. Thus, an equality cannot be obtained. If $F(0) \neq 0$, then Kay and Silverman have shown (Refs. B-4.6 and B-4.7), that we may write (B-69) more generally as

$$\Delta t \, \Delta\omega_+ > \frac{1}{2} \left| 1 - 2 |F(0)|^2 \omega_{o+} \right| \tag{B-71}$$

where ω_{o+} can be considered to be the centroid of the positive-frequency spectrum of $f(t)$.

The convolution of an input to a physical element with its impulse response is a degraded form of the input in the sense that any time-structure which is fine, compared with the element's "time constant," is smoothed out. The time-structure of any waveform may be expressed in terms of a temporal autocorrelation function; a measure of the smoothness of the waveform may be described by comparing its values at any two instants of time. It is also, to some extent, a measure of the "time constant" of the waveform.

Woodward (Ref. B-4.9), used the integral of the square of the normalized autocorrelation function as a measure of temporal extent. Taking absolute values (i.e., disregarding time structure), this is expressed as

$$\int_{-\infty}^{\infty} |\psi(\tau)|^2 d\tau = |r| . \tag{B-72}$$

When $f(t)$ and $f(t + \tau)$ become orthogonal or independent relative to each other, the correlation and consequently (B-72) vanishes. Since this occurs when τ is large and the constant

$|\tau|$ has the dimensions of time, $|\tau|$ may be considered to be a measure of the time-constant of the waveform, T. We then have

$$T = \int_{-\infty}^{\infty} |\psi(\tau)|^2 d\tau \qquad \text{(B-73)}$$

or equivalently

$$T = \int_{-\infty}^{\infty} |F(f)|^4 df \qquad \text{(B-74)}$$

where $F(f)$ is the Fourier spectrum of the waveform $f(t)$. Thus, T is a measure of the lack of orthogonality between the waveform and the same waveform displaced in time. Its reciprocal, $1/T$, is a measure of the frequency spread of $|F(f)|^2$.

Similarly, a "frequency constant," F, for any spectrum $F(f)$ may be written as

$$F = \int_{-\infty}^{\infty} |G(\phi)|^2 d\phi \qquad \text{(B-75)}$$

where $G(\phi)$ is the phase correlation function defined by (A-75). Equation (B-75) may also be expressed in the form

$$F = \int_{-\infty}^{\infty} |f(t)|^4 dt \,. \qquad \text{(B-76)}$$

Thus, F is a measure of the lack of orthogonality between the spectrum and the spectrum linearly displaced in frequency. Its reciprocal, $1/F$, is a measure of the extent to which $|f(t)|^2$ is spread out in time.

The product of the structural time constants T and F is

$$TF = \int_{-\infty}^{\infty} |\psi(\tau)|^2 d\tau \int_{-\infty}^{\infty} |G(\phi)|^2 d\phi \,. \qquad \text{(B-77)}$$

Changing the order of integration and combining the squared terms gives:

$$TF = \int_{-\infty}^{\infty}\int_{-\infty}^{\infty} |\psi(\tau)\, G(\phi)|^2 d\tau\, d\phi \,. \qquad \text{(B-78)}$$

However, we have found that

$$\psi(\tau,\phi) = \psi(\tau)\, G(\phi) \qquad \text{(B-79)}$$

if $f(t)$ is unmodulated. Substituting (B-79) into (B-78),

$$TF = \int_{-\infty}^{\infty}\int_{-\infty}^{\infty} |\psi(\tau,\phi)|^2 d\tau\, d\phi \qquad \text{(B-80)}$$

which, from (A-84) and (A-85), equals unity,

$$TF = 1 \,. \qquad \text{(B-81)}$$

In general, the product TF is not invariant for all waveforms since (B-78) is its general definition, and not (B-80). The latter is a special case. If the waveform is linearly frequency modulated, then

$$\psi(\tau,\phi) > \psi(\tau)\, G(\phi)\ . \tag{B-82}$$

Therefore, for the case of linear FM

$$TF << \int_{-\infty}^{\infty}\int_{-\infty}^{\infty} |\psi(\tau,\phi)|^2\, d\tau\, d\phi \tag{B-83}$$

$$TF << 1\ . \tag{B-84}$$

The product TF is a measure of the lack of orthogonality between the waveform and the same waveform displaced both in time and frequency. However, it is not an invariant measure as that formulated in (A-84) since it is usually dependent on the details of the waveform.

REFERENCES

B-4.1. R. E. Wernikoff, "Time-Limited and Band-Limited Functions," Quarterly Progress Report, Research Laboratory of Electronics, M.I.T., January 15, 1957

B-4.2. D. Gabor, "Theory of Information," Journ. Inst. Elect., Part III, Vol. 93 (1946)

B-4.3. R. M. Lerner, "The Representation of Signals," I.R.E., Trans., Vol. CT-6, pp. 197-216, May 1959

B-4.4. J. R. V. Olswald, "The Theory of Analytic Band-Limited Signals Applied to Carrier Systems," I.R.E. Trans., Vol. CT-3, pp. 244-251, December 1956

B-4.5. D. G. Lampard, "Definitions of 'Bandwidth' and 'Time Duration' of Signals Which are Connected by an Identity," I.R.E. Trans., Vol. CT-3, pp. 286-288, December 1956

B-4.6. I. Kay and R. A. Silverman, "On the Uncertainty Relations for Real Signals," Information and Control, Vol. 1, No. 1, pp. 64-75, September 1957

B-4.7. I. Kay and R. A. Silverman, "On the Uncertainty Relations for Real Signals: Postscript," Information and Control, Vol. 2, No. 4, pp. 396-397, December 1959

B-4.8. E. Wolf, "Reciprocity Inequalities, Coherence Time and Bandwidth in Signal Analysis and Optics," NYU Inst. Math. Sci., Dir. Electromagnetic Research, Res. Rept. No. EM-106

B-4.9. P. M. Woodward, "Time and Frequency Uncertainty in Waveform Analysis," Phil. Mag. Vol. 42, pp. 883-891, August 1951

C. EXAMPLES OF DESCRIPTIONS

Previous sections were concerned with methods of analysis available for representing functions, and the effect and importance of bounds on the representations. Bounds may be deterministic or statistical, or both — as they are for most problems in signal processing and their descriptions including the influence of bounds often constitutes the primary and the most difficult objective. The selection of a suitable representation for a signal may be made by considering the property to be characterized, and the use which is to be made of the representation. The fundamental concepts will be brought out by illustrations of deterministic, statistical, correlation and spectral descriptions of a few bounded and unbounded waveforms. Specifically, bounded and unbounded periodic waves and stationary random waveforms.

An infinite sine wave, Figure C-1(a), may be completely represented in the frequency domain as the sum of impulses at the positive and negative fundamental frequency, each containing half the power per period. Its autocorrelation function is periodic with the frequency of the sine wave and deletes all phase information, that is, if

$$x(t) = A \sin(\omega_o t + \phi) \tag{C-1}$$

then

$$\psi(\tau) = \frac{A^2}{2} \cos \omega_o \tau. \tag{C-2}$$

If the sine wave is bounded (Figure C-1(b)), consisting of a finite number of periods, the power is redistributed into major lobes at the fundamental frequencies and sidelobes. As the number of periods increases, the magnitude of the major lobes increases correspondingly and in the limit to an infinitely extended waveform, the Fourier spectrum will tend to become impulses. The autocorrelation function of the sine wave given by (C-1) bounded to $|t| \leq T$ may be expressed as

$$\begin{aligned} \psi(\tau) &= A^2 T \cos \omega_o \tau \left[1 - \frac{\sin 2\omega_o T}{2\omega_o T}\right] && \text{for } |\tau| \leq T \\ &= 0 && \text{for } |\tau| > T. \end{aligned} \tag{C-3}$$

This has the dimensions of energy while (C-2) has the dimensions of power. This is attributed to the unbounded waveform being "power bounded" whereas the bounded waveform is "energy bounded."

Although a unit impulse function (Figure C-1(c)) may appear to be bounded in time, it contains an infinite amount of energy, thereby implying an unbounded state. Its energy is proportional to the bandwidth and is therefore concentrated at the extremely high frequencies. The Fourier spectrum of the unit impulse has unit amplitude and zero phase for all frequencies. All the frequency components are in phase at $t = T$, which accounts for the height of the impulse at the specific instant of time.

If the unit impulse is repeated indefinitely at regular intervals (Figure C-1(d)), a train of impulses results having Fourier spectra also consisting of a train of impulses. Its autocorrelation function and power spectrum will be similar in form. The resultant waveform is now bounded in power.

The step function (Figure C-1(e)) is discontinuous at $t = 0$ and needs two specifications, for $t > 0$ and $t < 0$, to describe it in the time domain. The advantage of representing it in the

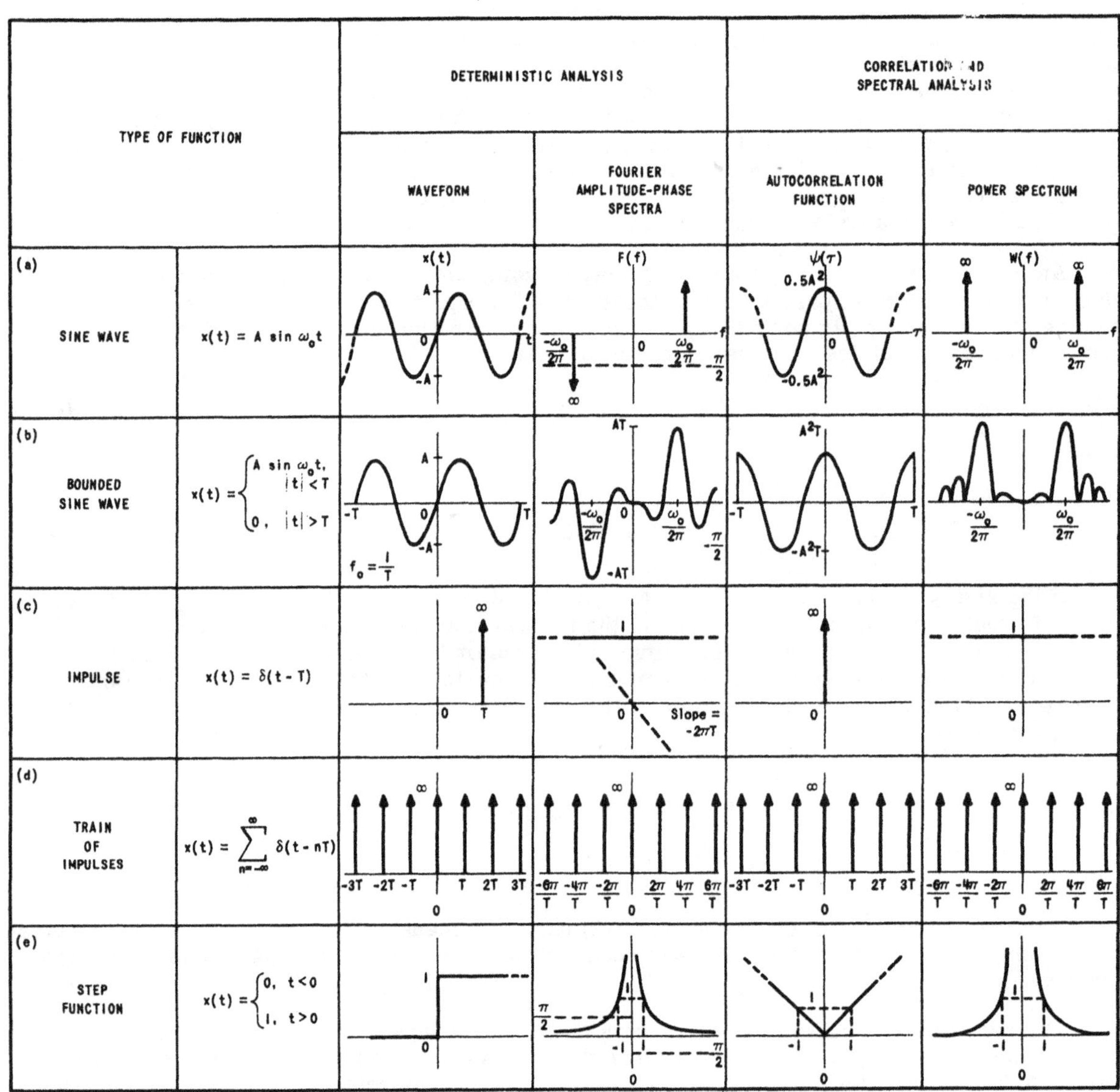

Figure C-1 (Continued) - Deterministic and correlation and spectral analysis of several bounded and unbounded waveforms

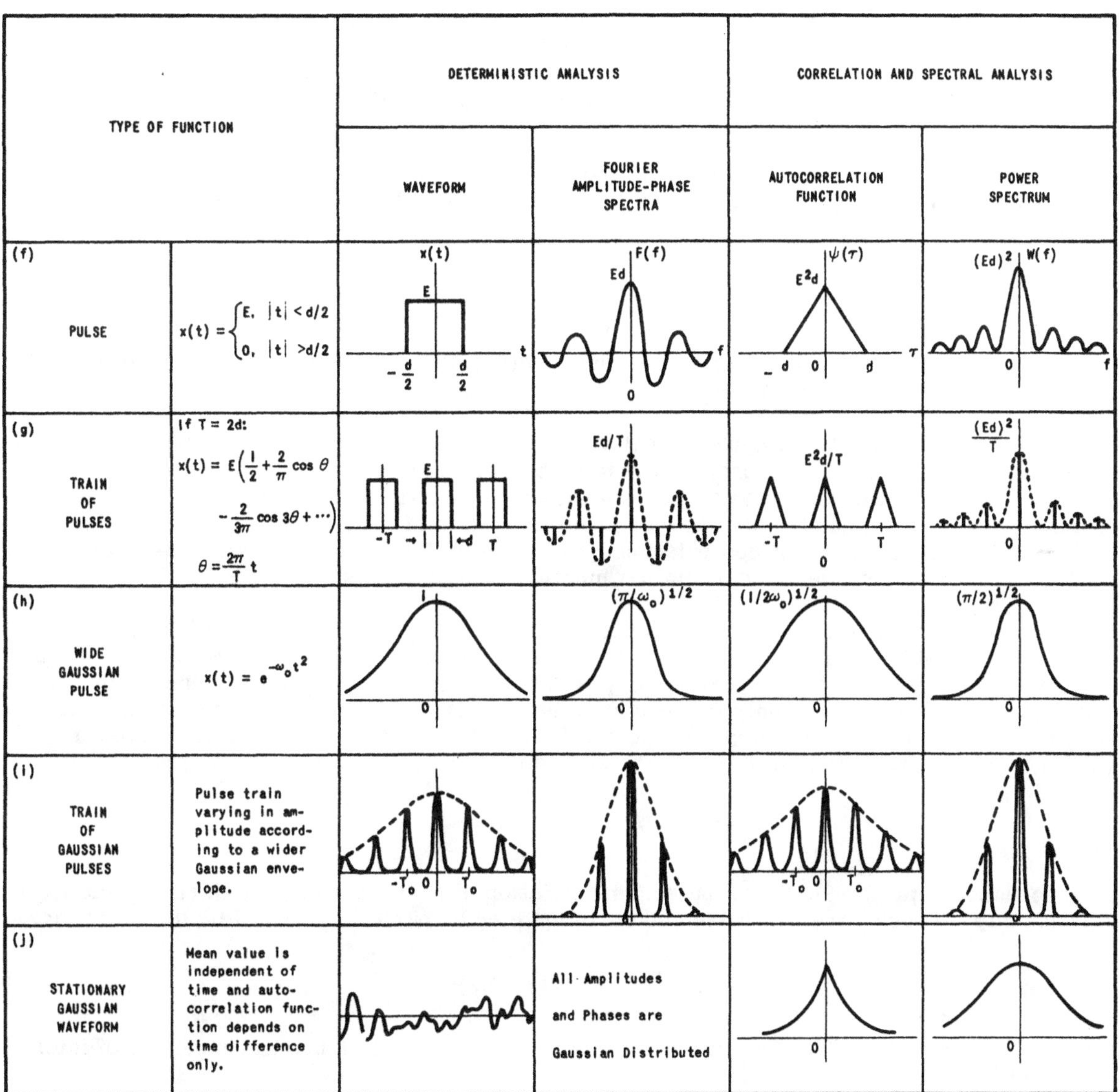

Figure C-1 - Deterministic and correlation and spectral analysis of several bounded and unbounded waveforms

frequency domain is that it has a continuous spectrum and hence, needs only a single specification for representation. Because of the concentration of high amplitude at the low end of the spectrum, low- and high-frequency effects are more equally depicted than for the unit impulse. This is a type of smoothing since the unit step is the integral of the unit impulse.

A pulse (Figure C-1(f)) has a sin x-over-x Fourier spectrum, which is the form of the response of a linear low-pass network to an impulse function. It can be considered to be a bounded step function, causing sidelobes to develop in the Fourier spectrum as they did for a bounded sine wave. The autocorrelation function is triangular having a width equal to the duration of the pulse and a continuous power spectrum having the form $\sin^2$ x-over-x^2, infinite in extent.

If pulses are repeated at regular intervals T (Figure C-1(g)) the autocorrelation function will also be periodic with period T. For a pulse height E and duration d, the autocorrelation function is given by

$$\psi(\tau) = \frac{E^2}{T}(d - |\tau + nT|) \tag{C-4}$$

where $|\tau + nT| \leq d$. The power spectrum is now discrete having components at integral values of $1/T$ and a $\sin^2$ x-over-x^2 envelope.

A wide Gaussian pulse (Figure C-1(h)) has a low-pass Gaussian Fourier spectrum. This illustrates the property of reciprocal spreading between conjugate Fourier descriptions. The power spectrum is also Gaussian and low-pass, and is of considerable importance because it simulates the gradual cutoff which is more representative of actual networks than abrupt transitions of idealized filters. A computational advantage of the Gaussian power spectrum is that its Fourier transform, the autocorrelation function, is also Gaussian.

A train of real positive Gaussian pulses is shown by Figure C-1(i), varying in amplitude according to a wider Gaussian envelope. Assuming that the pulses do not overlap and that a large number of them have comparable amplitudes, the forms of the Fourier conjugate descriptions are the same differing only in their parameters. If $(\Delta t)_e$ and $(\Delta\omega)_e$ are the time and frequency "widths" of the envelope, respectively, using the definition of duration given in (B-66), the time-bandwidth product for the envelope is

$$(\Delta t)_e \, (\Delta\omega)_e > \frac{1}{2} \tag{C-5}$$

in accordance with (B-69). For a pulse time duration of $(\Delta t)_p$ and bandwidth $(\Delta\omega)_p$, the interrelationship between Gaussian pulse and envelope may be partly expressed by the following:

$$(\Delta t)_e \, (\Delta\omega)_p = (\Delta t)_p \, (\Delta\omega)_e = \frac{1}{2} \tag{C-6}$$

The modulation has produced a lower limit for the time-bandwidth product measure of interdependency.

In general, the power spectrum, like the autocorrelation function, represents second-order statistics and does not give a complete description of the process. However, if the process has a Gaussian probability function, as in Figure C-1(j), statistics of all orders may be expressed in terms of the second order only. The autocorrelation function and power spectrum would then represent complete descriptions.

A deviation from the conventional indeterminate relationship between time duration and bandwidth is best depicted by the joint autocorrelation function. It is shown in Figure C-2 for the train of Gaussian pulses discussed above. The joint correlation function is made up of a lattice of elliptical Gaussian peaks having amplitudes which vary according to a Gaussian envelope indicated by the dotted contour. If T_o is the period of the pulses, the structural time constant (T) is

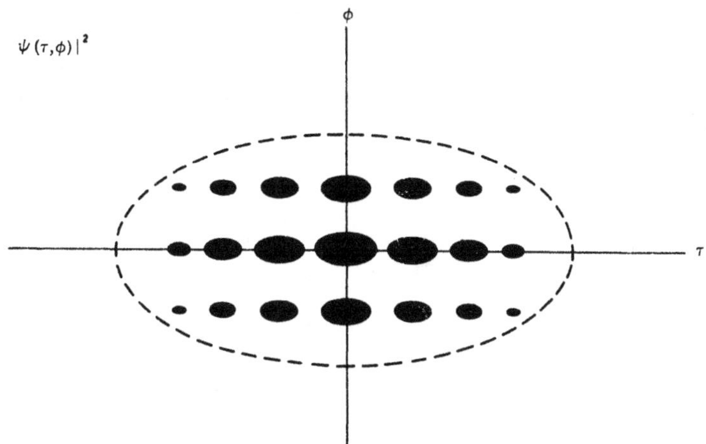

Figure C-2 - Squared envelope of joint autocorrelation function of a train of Gaussian pulses varying in amplitude according to a wider Gaussian pulse

$$T = 2\,\frac{(\Delta t)_e\,(\Delta t)_p}{T_o} \qquad \text{(C-7)}$$

and the frequency constant (F) is

$$F = 2\,T_o\,(\Delta f)_e\,(\Delta f)_p\,. \qquad \text{(C-8)}$$

Using (C-6), the product of T and F is found to equal unity,

$$TF = 1\,. \qquad \text{(C-9)}$$

This is attributed to the separability of the joint autocorrelation function of Figure C-2.

Statistical descriptions exist for the waveforms above but they do not simplify the representations. Figure C-3 shows the probability density function and characteristic function for an infinite sine wave, a train of pulses, and a stationary Gaussian process. The density function for the sine wave (C-1) was obtained for t fixed, the random variable θ uniformly distributed from $-\pi$ to π, and is given as

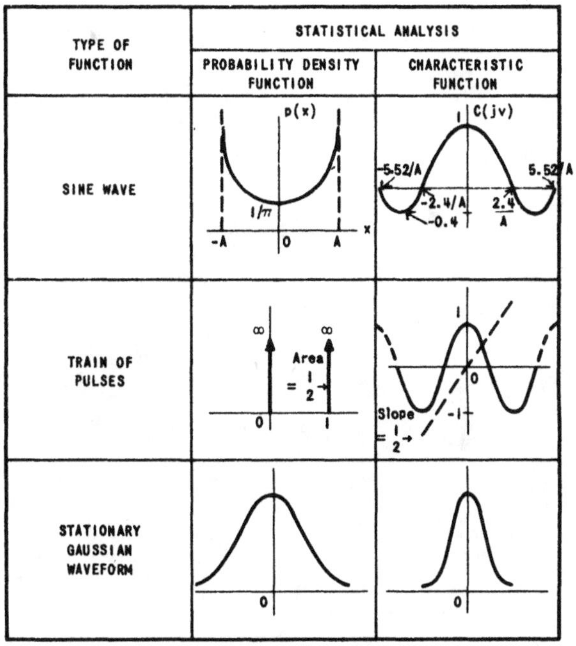

TYPE OF FUNCTION	STATISTICAL ANALYSIS	
	PROBABILITY DENSITY FUNCTION	CHARACTERISTIC FUNCTION
SINE WAVE	p(x); 1/π; -A, 0, A; x	C(jv); 1; -5.52/A, 5.52/A; -2.4/A, 2.4/A; -0.4
TRAIN OF PULSES	∞, ∞; Area = 1/2; 0, 1	1; 0; Slope = 1/2; -1
STATIONARY GAUSSIAN WAVEFORM	0	0

Figure C-3 - Statistical analysis for several waveforms

$$p(x) = \left.\begin{array}{ll} \dfrac{1}{\pi\,(A^2 - x^2)^{1/2}} & \text{for } |x| \le A \\ = 0 & \text{for } |x| > A \end{array}\right\}. \qquad \text{(C-10)}$$

For the train of pulses, there are only two values equally probable, x = 0 and x = 1, so that the density function is concentrated at these points in the form of impulses, each of area 1/2,

$$p(x) = \frac{1}{2}\,\delta(x) + \frac{1}{2}\,\delta(x-1). \qquad \text{(C-11)}$$

The density function for a stationary Gaussian waveform is clearly Gaussian and possesses the properties peculiar to such functions. For the waveforms illustrated, correlation and spectral analysis may be adequate.

D. ADDITIONAL DESCRIPTIONS

1. INTRODUCTION

The imposition of bounds on functions has led to the concept of sampling which permitted specifying the functions by discrete values. The discrete values need not necessarily be values of the function. They can be values of other significant parameters such as derivatives of various orders or integrals. Too, the samples need not be uniformly distributed. However, their use would require knowledge not only of magnitudes but also of the time instants at which they were obtained.

2. DERIVATIVE SAMPLING

One extension of the sampling theorem permits the determining of the periodic sampling interval when the instantaneous sampling includes the amplitude and derivative values.

When the first derivative alone is added to the function amplitude sample, the sampling interval is $T_s = 1/W$, which is twice the interval required when only amplitude samples are made. Addition of each succeeding derivative allows the time interval between samples to become larger according to $T_s = (K+1)/2W$ where K is the order of the highest derivative when all lower ordered derivatives are observed in each sample.

The sampling theorem may be stated as follows: if a function $f(t)$ contains no frequency higher than W cps, it is determined by giving M function derivate values at each of a series of points extending throughout the time domain. The sampling interval $T_s = M/2W$ is then the interval between instantaneous observations. The recovery formula when the derivative values are included becomes increasingly complex. For the case when only the function and its first derivative are considered, the equation becomes:

$$f(t) = \sum_{n=-\infty}^{\infty} \left[f(n/W) + (t - n/W)\, f'(n/W)\right] \left[\frac{\sin \pi(Wt - n)}{\pi(Wt - n)}\right]^2 \qquad \text{(D-1)}$$

The composing function and its transform is illustrated in Figure B-1(d).

It should be pointed out that this does not conflict with the previous statement that $2WT$ sample values are required to specify a function of duration T and bandwidth W. Actually, it indicates another method by which $2WT$ independent samples may be obtained.

Consider the case where N equals $N/2$ amplitudes and $N/2$ first derivatives of the signal. A Fourier analysis will then yield the amplitudes of $N/4$ sine terms and $N/4$ cosine terms so that all harmonics up to the $(N/4)$th will be known. In order for a channel to pass all harmonics up to $(N/4T)$, its bandwidth $(BW)'$ must be at least

$$(BW)' \geq \frac{N}{4T}. \qquad \text{(D-2)}$$

Since it is assumed the N symbols still contain the same information as when N represented amplitude values,

$$N = 4T\,(BW)' = 2WT \qquad \text{(D-3)}$$

$$(BW)' = \frac{W}{2}\,. \qquad \text{(D-4)}$$

The Shannon sampling interval, T_s, is then equal to

$$T_s = \frac{1}{2(BW)'} = \frac{1}{W}\,. \qquad \text{(D-5)}$$

The same reasoning is valid when the N symbols are comprised of

N/3 amplitudes = n/6 sine terms plus n/6 cosine terms
+
N/3 first derivatives
+
N/3 second derivatives

The effective bandwidth $(BW)''$ is given as

$$(BW)'' = \frac{N}{6T} \qquad \text{(D-6)}$$

$$N = 6T(BW)'' = 2WT \qquad \text{(D-7)}$$

$$(BW)'' = \frac{W}{3}\,. \qquad \text{(D-8)}$$

Hence, when the second derivatives of the function are considered, the sampling interval may be expressed as

$$T_s = \frac{1}{2(BW)''} = \frac{3}{2W}\,. \qquad \text{(D-9)}$$

These results are tabulated below:

K = Order of derivative	0	1	2
Sampling interval	1/2W	2/2W	3/2W

and leads inductively to the following formula:

$$\text{Sampling Interval} = \frac{K+1}{2W}\,. \qquad \text{(D-10)}$$

3. NONUNIFORM SAMPLING

NONUNIFORM AMPLITUDE SAMPLING

It was mentioned earlier that a number of approximations can be used to represent a given function. One type of approximation is the Lagrangian interpolation polynomial whose values coincide with those of the given function at a specified number of points. Any polynomial of the nth degree is exactly specified by $n + 1$ points and has n zeros (including multiplicity of zeros). The Lagrangian polynomial has the form

$$f_n(t) = f_o L_o^n(t) + f_1 L_1^n(t) + \cdots + f_n L_n^n(t) \qquad \text{(D-11)}$$

where $L_j^n(t)$ is the Lagrangian coefficient defined by

$$L_j^n(t) = \frac{(t-t_o)(t-t_1)\cdots(t-t_{j-1})(t-t_{j+1})\cdots(t-t_n)}{(t_j-t_o)(t_j-t_1)\cdots(t_j-t_{j-1})(t_j-t_{j+1})\cdots(t_j-t_n)} . \tag{D-12}$$

This has the property:

$$L_j^n(t_i) = \begin{cases} 0, & i \neq j \\ 1, & i = j \end{cases} \tag{D-13}$$

so that each sample point in time will correspond to only one term in (D-11) which will have a coefficient of unity. Thus, it can be seen that the entire polynomial will agree with the sample data at each of the sample points. This interpolation polynomial which yields f_j at $t = t_j$ for $0 \leq j \leq n$ can be put in the form of (D-11).

If $g_n(t)$ is defined as

$$g_n(t) = \prod_{n=-\infty}^{\infty} \left(1 - \frac{t}{t_n}\right) \tag{D-14}$$

then the Lagrangian coefficient can be expressed as

$$L_n(t) = \frac{g_n(t)}{(t-t_n)\, g_n'(t_n)} \tag{D-15}$$

Making the Lagrangian interpolation polynomial more inclusive, (D-11) becomes

$$f_n(t) = g_n(t) \sum_{n=-\infty}^{\infty} \frac{f(t_n)}{(t-t_n)\, g_n'(t_n)} \tag{D-16}$$

$$= \sum_{n=-\infty}^{\infty} f(t_n)\, L_n(t). \tag{D-17}$$

This is quite similar to the reconstruction formula when employing uniform sampling, i.e.,

$$f(t) = \sum_{n=-\infty}^{\infty} f(t_n)\, \phi_n(t) \tag{D-18}$$

$$= \sum_{n=-\infty}^{\infty} f\left(\frac{n}{2W}\right) \frac{\sin \pi(2Wt - n)}{\pi(2Wt - n)} . \tag{D-19}$$

Therefore, (D-17) can be considered to be the form of a general sampling theorem for arbitrary sampling instants t_n which reduces to (D-19) as a special case for $t_n = n/2W$. Consequently, for systems dealing with discrete signals occurring at irregular intervals, one may employ a nonuniform amplitude sampling where the sampling function is expressed in terms of the Lagrange interpolation functions. Although the instants of sampling t_n are arbitrary, the average spacing between successive instants is $1/2W$. In all sampling schemes the average rate of sampling cannot be less than $2W$ per second (the Nyquist rate).

Figure D-1 illustrates six methods of sampling over a finite time interval having a duration T. The first (a) is uniform amplitude sampling and the remaining five (b-f) depict derivative sampling of increasing order (assuming that the derivatives exist at the point of evaluation). It is seen that for the size of the sampling interval ($1/2W$) chosen, sampling processes (a), (b), (c), (f), are uniform, where only the value of the amplitude or derivatives of the samples need be known. In process (d) four evaluations at t_4 are made. To complete the necessary

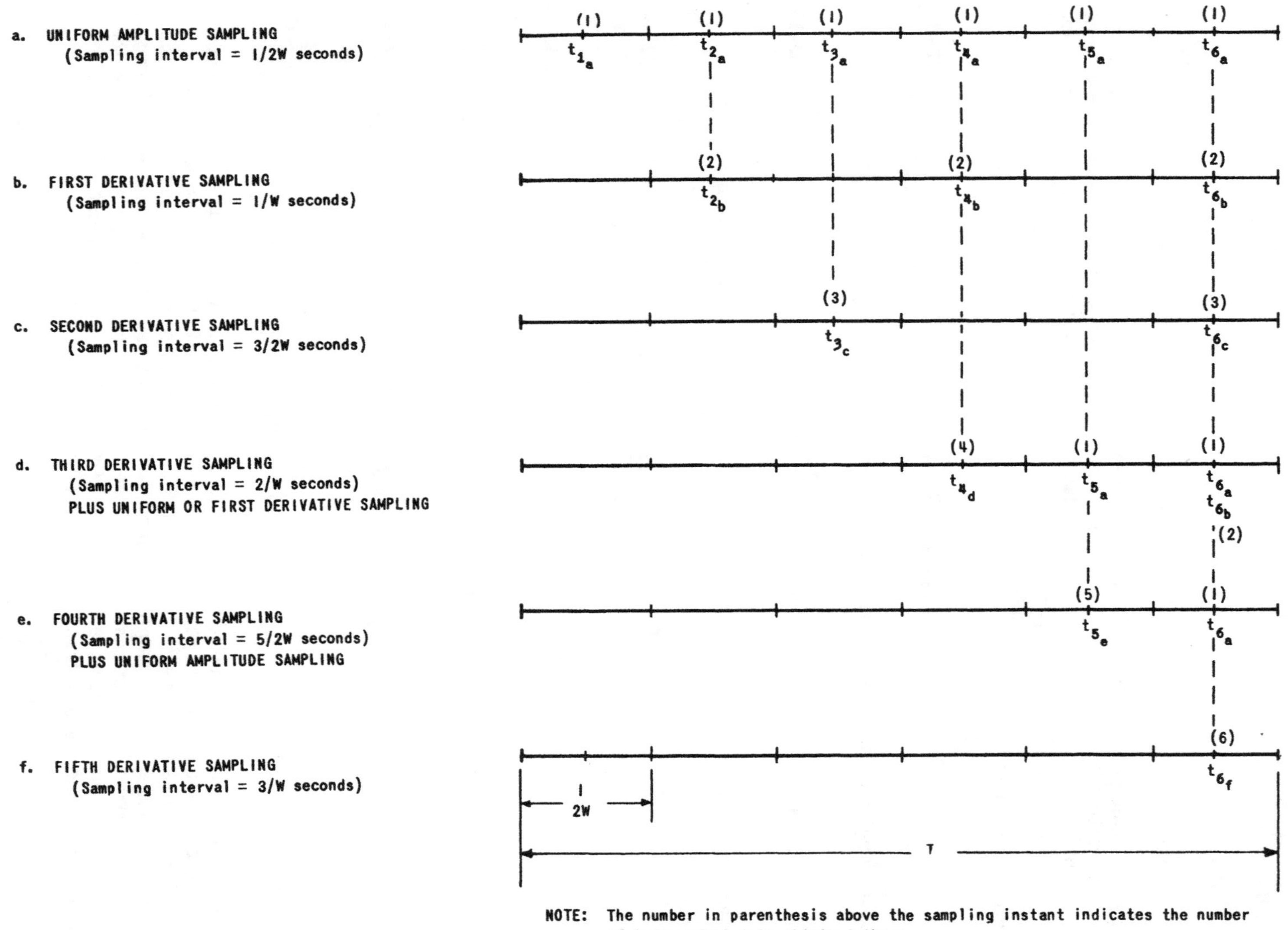

Figure D-1 - Six methods of sampling over a finite time interval having a duration T

amount of data we can sample at $1/2W$ for the remainder of the interval and obtain the sample values (amplitude) received in process (a) at instants t_5 and t_6 or sample at $1/W$ and obtain the two sample values (amplitude and first derivative) received in process (b) at instant t_6. No matter which sampling technique is used to complete the evaluation of the function in the interval, because more than one technique was used for a complete analysis resulting in non-uniform sampling intervals, it is necessary to have a knowledge of not only the magnitude of the $2WT$ sample points, but also the instants at which they were taken. Then, and only then, can the signal be recovered. A similar analysis can be applied to process (e).

Reference (D-3) indicates that the interpolation functions, $L_n(t)$, are bandlimited as long as the number of nonuniform intervals are finite, and derives four generalized theorems for describing the nonuniform sampling of bandlimited signals.

Thorem I: Migration of a Finite Number of Uniform Sample Points (See Figure D-2)

If a finite number of uniform sample points in a uniform distribution are migrated to new distinct positions $t = t_p$ thus forming a new distribution denoted by $t = \tau_m$, the bandlimited signal $f(t)$ will be uniquely defined. When N uniform sample points located at $t = n_p/2W$, where n_p with $p = 1, 2, \ldots N$ are N distinct integers, are migrated to N new positions $t = t_p$, $2Wt_p$ is not an integer.

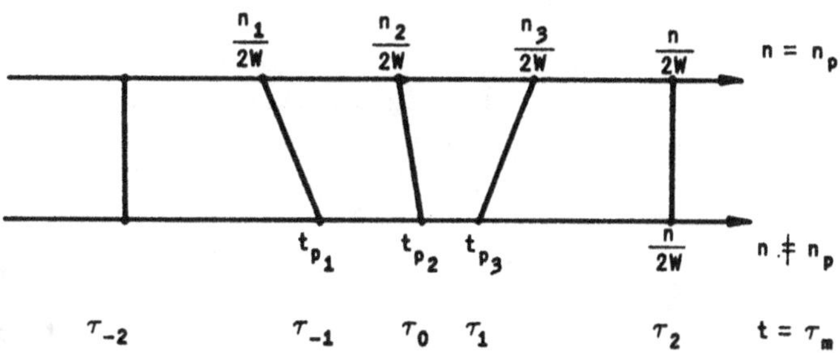

Figure D-2 - Sample point distribution for Theorem I

Theorem II: Sampling with a Single Gap in an Otherwise Uniform Distribution (See Figure D-3)

When the number of shifted uniform sample points increase without bound, Theorem I is no longer valid and the reconstruction function will generally become extremely complicated. A special case which is simpler in analysis is if half of the uniform sample points, say all those with $t > 0$, are shifted by an equal amount with respect to the rest. All sample points may then be denoted by τ_m with $\tau_m = \Delta t + p/2W$ where $p = 0, 1, \ldots$. For such a distribution, if the gap Δt is positive and less than $1/2W$ the signal is uniquely specified.

This theorem illustrates the effect that a particular determination of the sample points of a signal has upon the reconstruction. When $0 < \Delta t < 1/2W$, the signal can be uniquely reconstructed. It should be noted that when the number of samples are finite and equal to $2WT$ the reconstruction series will reduce to one having a sine-over-sine composing function. If $1/2W < \Delta t < 2/2W$, then the signal will be determined except for one arbitrary constant; this process is known as underspecification. If $-1/2W < \Delta t < 0$, then the sample values cannot be arbitrarily assigned but must satisfy a consistency condition; this is known as overspecification.

Theorem III: Recurrent Nonuniform Sampling (See Figure D-4)

If the sample points are divided into groups of N points each, having a recurrent period of $N/2W$ seconds, in one period the points may be denoted by t_p, $p = 1, 2, \ldots N$. For such a

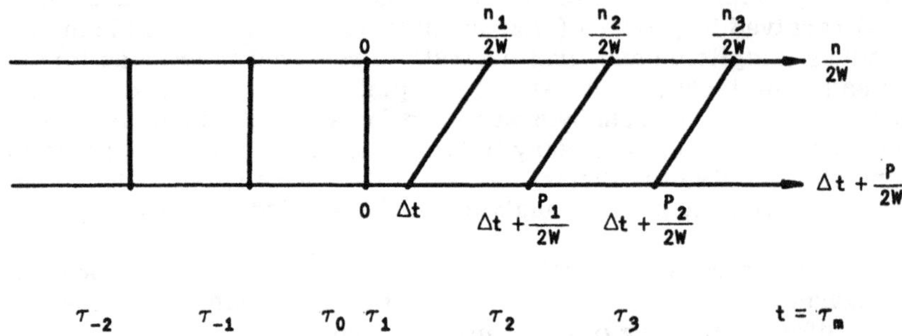

Figure D-3 - Sample point distribution for Theorem II

distribution, a bandlimited signal is uniquely determined by its values at a set of recurrent sample points $t = \tau_{pm} = t_p + mN/2W$, $p = 1, 2, \ldots, N$; $m = \ldots, -1, 0, 1, \ldots$ (m denotes group designation).

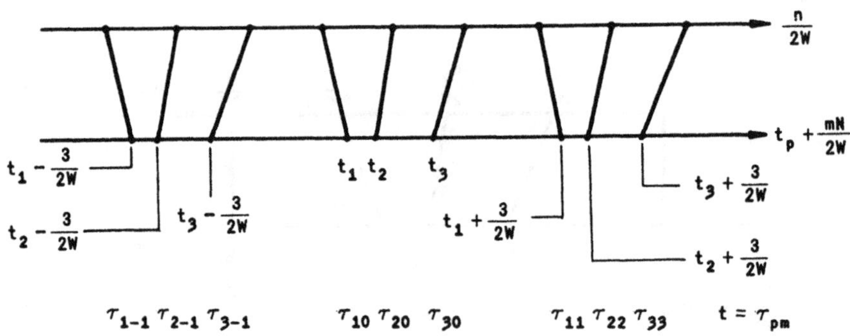

Figure D-4 - Sample point distribution for Theorem III

Theorem IV: "Minimum-Energy" Signals

A time-limited (T) signal of finite bandwidth (W) may be specified by 2WT equally spaced samples or by 2WT arbitrarily distributed samples using Theorem I. If we do not wish to specify the time interval, 2WT arbitrarily distributed samples can be used to uniquely define a "minimum-energy" signal, that is, a signal $f(t)$ with no frequency component above W cps whose energy

$$\int_{-\infty}^{\infty} f(t)^2 \, dt$$

is a minimum. The time interval inferred is that corresponding to the passage of this minimum energy. When the sample points are taken from a uniform distribution, the time-limited and minimum energy signals become identical.

From the reconstruction functions for the theorems given above, formulated in Table D-1, it is seen that for a nonuniform distribution composing functions for different sample points do not have the same form. Also, as the sample point deviates more and more from the uniform one, the composing functions become increasingly complicated. For a uniform distribution the maximum value of a composing function for a particular sample point occurs at the sample point and is unity. In a nonuniform distribution, the value of the composing function at its

TABLE D-1
Nonuniform Sampling Methods

NONUNIFORM SAMPLING	SAMPLING THEOREM	RECONSTRUCTION FUNCTION $f(t) = \sum_{m=-\infty}^{\infty} f(t_n)\, L_n(t)$
I. MIGRATION OF A FINITE NUMBER OF UNIFORM SAMPLE POINTS (SEE FIGURE D-2)	If a finite number of uniform sample points in a uniform distribution are migrated to new distinct positions $t = t_p$ thus forming a new distribution denoted by $t = t_m$, the bandwidth-limited signal $f(t)$ will be uniquely defined. When N uniform sample points located at $t = n_p/2W$, where n_p with $p = 1,2,\ldots,N$ are N distinct integers, are migrated to N new positions $t = t_p$, $2Wt_p$ is not an integer. The reconstruction formula will have the form $$f(t) = \sum_{m=-\infty}^{\infty} f(\tau_m)\, \psi_m(t)$$	$$f(t) = \sum_{m=-\infty}^{\infty} f(\tau_m)\, \psi_m(t)$$ $$\psi_m(t) = \frac{\prod_{q=1}^{N} (t - t_q) \prod_{q=1}^{N}\left(\frac{n}{2W} - \frac{n_q}{2W}\right)}{\prod_{q=1}^{N}\left(t - \frac{n_q}{2W}\right)\prod_{q=1}^{N}\left(\frac{n}{2W} - t_q\right)} \cdot \frac{(-1)^n \sin 2\pi Wt}{\pi(2Wt - n)} \qquad \psi_m(t) = \frac{\prod_{q=1, \neq p}^{N} (t - t_q) \prod_{q=1}^{N}\left(t_p - \frac{n_q}{2W}\right)}{\prod_{q=1}^{N}\left(t - \frac{n_q}{2W}\right)\prod_{q=1, \neq p}^{N}(t_p - t_q)} \cdot \frac{\sin 2\pi Wt}{\sin 2\pi Wt_p}$$ for $\tau_m = t_p$ where $2Wt_p$ is not an integer
II. SAMPLING WITH A SINGLE GAP IN AN OTHERWISE UNIFORM DISTRIBUTION (SEE FIGURE D-3)	If half of the uniform sample points, say all those with $t > 0$, are shifted by an equal amount with respect to the rest, all sample points may then be denoted by τ_m with $\tau_m = -p/2W$ or $\tau_m = \Delta t + p/2W$ where $p = 0,1,\ldots$ For such a distribution, if the gap Δt is positive and less than $1/2W$ the signal is uniquely specified. The reconstruction formula will have the form of that of Theorem I.	$$f(t) = \sum_{m=-\infty}^{\infty} f(\tau_m)\, \psi_m(t)$$ $$\psi_m(t) = \frac{(-1)_p\, \Gamma(2W\Delta t + p)}{\Gamma(2Wt)\, \Gamma[2W(\Delta t - t)]\, p!}$$ $$\frac{1}{p + 2Wt}, \quad \tau_m = -\frac{p}{2W}$$ $$\frac{1}{pp + 2W(\Delta t - t)}, \quad \tau_m = \Delta t + \frac{p}{2W};\ p = 0,1,2,\ldots$$ OVERSPECIFICATION: $(M-1)/2W < \Delta t < M/2W$ UNDERSPECIFICATION: $M/2W < \Delta t < (M+1)/2W$ where m is a positive integer
III. RECURRENT NONUNIFORM SAMPLING (SEE FIGURE D-4)	If the sample points are divided into groups of N points each, having a recurrent period of N/2W seconds, in one period the points may be denoted by t_p, $p = 1,2,\ldots N$. For such a distribution, a bandwidth-limited signal is uniquely determined by its values at a set of recurrent sample points $t = \tau_{pm} = t_p + mN/2W$, $p = 1,2,\ldots N$; $m = \ldots, -1, 0, 1,\ldots$ (m denotes group designation). The reconstruction is $$f(t) = \sum_{m=-\infty}^{\infty} \sum_{p=1}^{N} f(\tau_{pm})\, \psi_{pm}(t)$$	$$f(t) = \sum_{m=-\infty}^{\infty} \sum_{p=1}^{N} f(\tau_{pm})\, \psi_{pm}(t)$$ $$\psi_{pm}(t) = \frac{\prod_{q=1}^{N} \sin \frac{2\pi W}{N}(t - t_q)}{\prod_{q=1, \neq p}^{N} \sin \frac{2\pi W}{N}(t_p - t_q)} \cdot \frac{(-1)^{mN}}{\frac{2\pi W}{N}\left(t - t_p - \frac{mN}{2W}\right)}$$
IV. "MINIMUM-ENERGY" SIGNALS	If the sample values at a finite set of arbitrarily distributed sample points $t = \tau_p$, $p = 1,2,\ldots,N$ are given, a signal $f(t)$ with no frequency component above W cps is defined uniquely under the condition that the "energy" of the signal $$\int_{-\infty}^{\infty} f(t)^2\, dt$$ is a minimum. The reconstruction formula of the signal is $$f(t) = \sum_{p=1}^{N} f(\tau_p)\, \psi_p(t)$$	$$f(t) = \sum_{p=1}^{N} f(\tau_p)\, \psi_p(t)$$ $$\psi_p(t) = \sum_{q=1}^{N} A_{qp} \frac{\sin 2\pi W(t - \tau_q)}{2\pi W(t - \tau_q)}$$ The coefficients A_{qp} are those of the inverse matrix of one whose elements are $$\frac{\sin 2\pi W(\tau_p - \tau_q)}{2\pi W(\tau_p - \tau_q)} \qquad p,\ q = 1,2,\ldots N$$

particular sample point remains unity, although this may not be its maximum value. As sample points are bunched closer together, the maximum value of composing functions of sample points close to the gap produced by the bunching will tend to become larger. This can be seen in Figure D-5 which depicts a composing function for the sample point distribution corresponding to Theorem I.

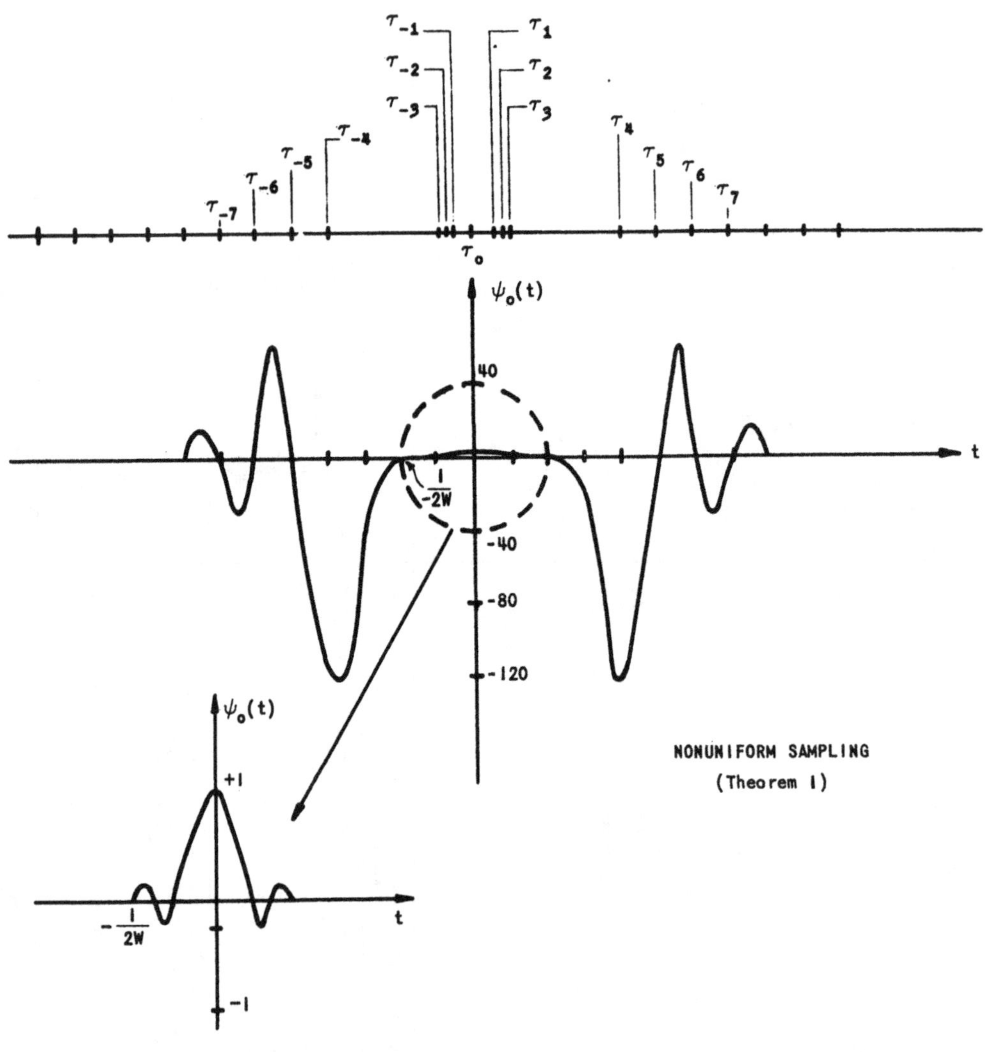

Figure D-5 - A composing function for the sample point distribution τ_m obtained from the migration of uniform sample points $t = \pm 2/2W$ and $t = \pm 3/2W$ to $t = \pm 1/4W$ and $t = \pm 3/8W$

When two adjacent sample points are bunched together, the main information is contained in the values of the signal and its derivative near the two sample points. Since the maximum values of the derivative is needed, this implies that a greater accuracy is needed in determining the sample values for an accurate reconstruction of the function. Consequently, for physical systems with limited dynamic range, the total number of distinguishable signals in an observation process employing nonuniform sampling may decrease even though the degrees-of-freedom are preserved.

In view of what has been said, a general sampling theorem may now be given as follows:

If a signal is a magnitude-time function, and if time is divided into equal intervals of T seconds where $T = N/2W$, and if N instantaneous samples are taken from each interval in any manner, then a knowledge of the instant at which the sample is taken determines the original signal uniquely.

A sampled wave can be represented by any set of 2WT independent numbers associated with the function, and these represent the least number of values capable of completely and uniquely defining the function. This includes derivatives and integrals. The total number necessary per period is fixed and need not be equally spaced. If the independent numbers are bunched to a substantial extent, the values must be known with extraordinary precision to afford accurate reconstruction of the function.

SAMPLING THE ZEROS OF BANDLIMITED SIGNALS

One form of nonuniform sampling is the sampling of the zeros of bandlimited signals. In this method, the sampling points are determined by the characteristics of the signal containing the message. Information is transmitted over a channel by preserving the occurrence of zero crossings rather than denoting amplitudes or slopes at specified instants.

In general, the average rate of zero crossings of a bandlimited signal is less than the Nyquist rate. However, the use of high-order derivatives of the signal will result in a waveform whose zeros approach the Nyquist rate though they will be very closely correlated and will no longer represent independent samples. A continuous bandlimited function will include "complex conjugate" zeros which, unlike real zeros, are not physically detectable, and will tend to cluster along the real axis.

The above considerations leads to the following formula which gives the synthesis of a bandlimited function with a given set of real and complex zeros within an interval, assuming real zeros at the Nyquist rate outside the interval:

$$f(t) = \sum_{n=-N}^{N} (-1)^n A_n \frac{\sin \pi (2Wt - n)}{\pi (2Wt - n)} \qquad \text{(D-20)}$$

where:

$$A_n = f(0) \frac{\prod_m \frac{1}{2\pi Z_m} \prod_m (2WZ_m - n)}{\prod_{m=1}^{N} - \left(\frac{1}{m}\right)^2 \prod_{\substack{m=-N \\ m \neq n}}^{N} (m - n)} \qquad \text{(D-21)}$$

Z_m = complex zeros = $t_m + ju_m$

N = integer not exceeding WT ($N \leq WT$).

This can be seen to be very similar to the uniform sampling formula for a bandlimited function.

The amplitudes at the sampling points (A_n) can be expressed in terms of the migrations of the zeros from the uniform sample point locations. The results show that the location of a zero (or migration from a sample point location) affects the amplitude of the signal in its immediate vicinity but does not have a marked effect on the signal at a much earlier or later time. A large migration, resulting in a large interval between successive zeros, will produce a large signal amplitude.

A binary signal, such as infinitely clipped speech, can be replaced by a bandlimited signal having the same zero crossings. If the complex zeros are included, the bandwidth is increased and the signal amplitude reduced from that which would have resulted from the real zeros. This illustrates how bandwidth may be exchanged for signal-to-noise ratio when information is transmitted. An illustration of the theory given is shown in Figure D-6.

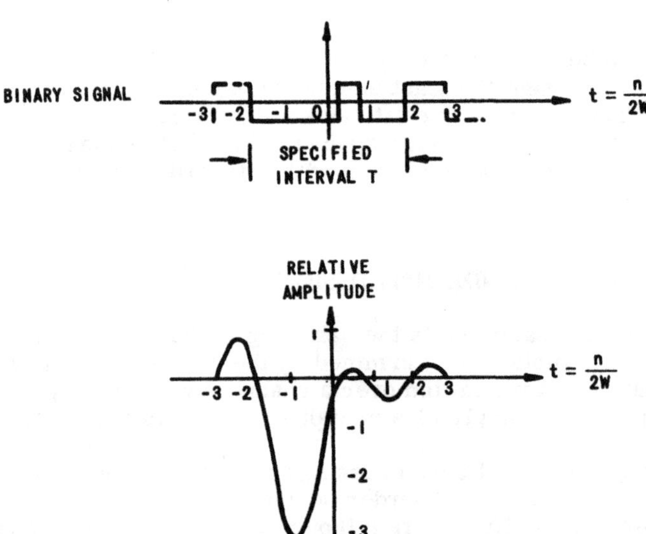

Figure D-6 - Bandlimited signal having specified zero crossings

The abrupt crossings of a binary signal can be transmitted over a channel by converting the migration intervals to pulse amplitudes occurring at uniform intervals 1/2W with the individual amplitudes proportional to the corresponding migration. This would give 2W average numbers of crossings per second with the minimum theoretical transmission bandwidth remaining at W. It should be emphasized that the information that can be recovered is dependent upon the dynamic range of the system which, as previously mentioned, is proportional to the distribution of the intervals between crossings. These principles are illustrated in Figure D-7.

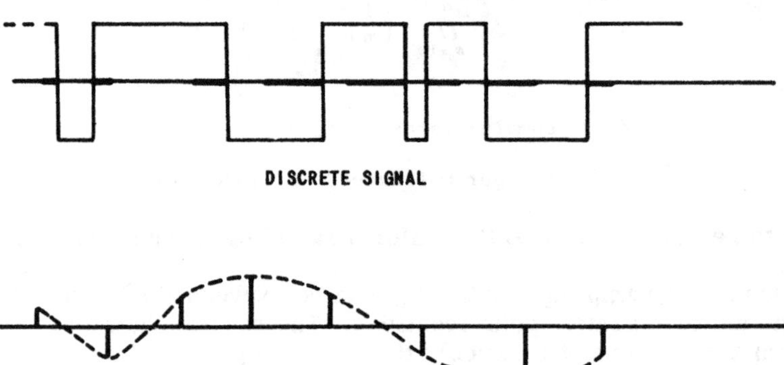

Figure D-7 - Conversion of the migration of zero crossings of a discrete signal to pulse amplitude at uniform intervals

4. AMPLITUDE QUANTIZATION

A continuous signal with a finite amplitude range will have an infinite number of amplitude levels. It is not possible or necessary to transmit the exact amplitudes of the samples. Consider the sample in Figure D-8. A signal may be transmitted with a finite number of discrete amplitudes if all samples such as OM can be considered equal when M lies within the amplitude range q. It is then permissible to represent and transmit all amplitude levels within this range by one discrete amplitude ON. The signal recovered will be different from the original but since the maximum error cannot exceed one-half step, the deviation from fidelity can be reduced by increasing the number of quantum states or amplitude levels, keeping the total amplitude range constant.

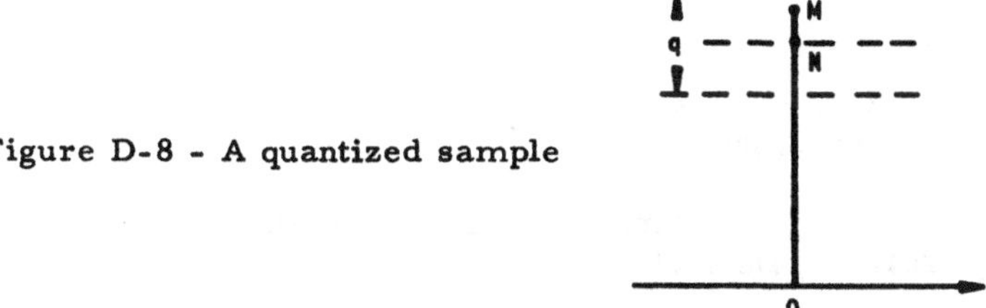

Figure D-8 - A quantized sample

Representing the signal by certain discrete allowable levels only is called quantizing. It inherently introduces an initial error in the amplitude of the samples, giving rise to quantization noise. This is the difference in signal power before and after quantizing. Quantization is a nonlinear operation which occurs whenever a continuous physical process is represented numerically. Use of quantization within a system may complicate analysis. However, there are methods which may reduce the complexity for some applications.

A quantizer is a device which processes continuous data or sampled data. It has the property that an input lying somewhere within a quantization "box" of width q will yield an output corresponding to the center of that box. The input-output characteristic of a quantizer is illustrated in Figure D-9. The probability density of the output, $p'(x)$, will consist of a series of impulses that are uniformly spaced along the amplitude axis. Each impulse will be centered in a quantization box and have a magnitude equal to the area under the probability density $p(x)$ within the bounds of the box. The quantizer output distribution $p'(x)$ consists of "area samples" of the input density $p(x)$ and the quantizer may consequently be thought of as an area sampler. Therefore, amplitude quantization may be considered to be a sampling of the probability density of the functions in question.

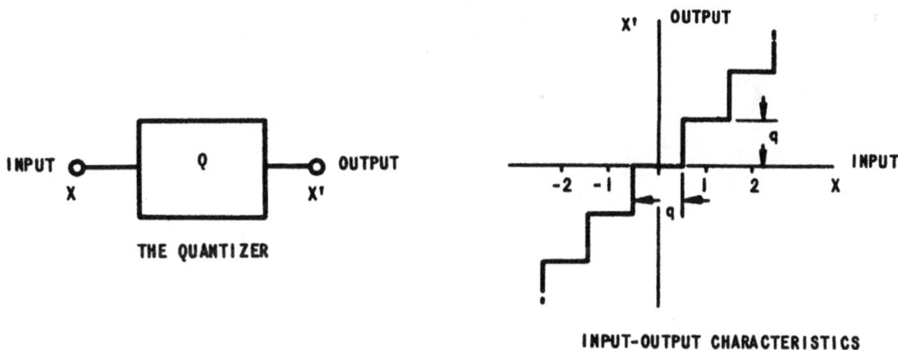

Figure D-9 - The quantizer and its input-output characteristics

The probability density of quantization noise Q_n is independent of the probability density of the input to the quantizer as long as the Shannon criterion is satisfied. This implies that when the radian "fineness" $\phi = 2\pi/q$ is twice the radian "frequency" of the highest "frequency" component contained in $p(x)$, $p(x)$ can be completely recovered from the quantized density $p'(x)$. The distribution of noise introduced by the quantizer will then be flat-topped having a sin x-over-x characteristic function.

Quantization can be considered as a sampling process that acts not upon the function itself but upon its probability density. As in conventional sampling, where the Shannon criterion allows a function to be recovered, a sampling theorem for quantization exists such that if the quantization is sufficiently fine, the statistics are recoverable.

REFERENCES

D-1. L. J. Fogel, "A Note on the Sampling Theorem," I.R.E. Trans., Vol. IT-1, pp. 47-48, March 1955

D-2. D. L. Jagerman and L. F. Fogel, "Some General Aspects of the Sampling Theorem," I.R.E. Trans., Vol. IT-2, pp. 139-145, December 1956

D-3. J. L. Yen, "On Non-Uniform Sampling of Bandwidth-Limited Signals," I.R.E. Trans., Vol. CT-3, pp. 251-257, December 1956

D-4. F. E. Bond and C. R. Cahn, "On Sampling the Zeros of Bandwidth Limited Signals," I.R.E. Trans., Vol. IT-4, pp. 110-113, September 1958

D-5. B. Widrow, "A Study of Rough Amplitude Quantization by Means of Nyquist Sampling Theory," I.R.E. Trans., Vol. IT-2, pp. 266-276, December 1956

D-6. W. R. Bennett, "Spectra of Quantized Signals," Bell System Technical Journal, Vol. 27, pp. 446-472, July 1948

D-7. H. S. Shapiro and R. A. Silverman, "Alias-Free Sampling of Random Noise," J. Soc. for Indust. and Appl. Math., Vol. 8, No. 2, pp. 225-248, June 1960

D-8. A. Kohlenberg, "Exact Interpolation of Band-Limited Functions," J. Appl. Phys., Vol. 24, pp. 1432-1436, December 1955

D-9. D. Middleton, "A Comparison of Random and Periodic Data Sampling for the Detection of Signals in Noise," I.R.E. Trans., Vol. CT-6, Special Supplement, pp. 234-247, May 1959

D-10. D. Middleton and D. Van Meter, "Detection and Extraction of Signals in Noise from the Point of View of Statistical Decision Theory, I and II," J. Soc. for Indust. and Appl. Math., Vol. 3, No. 4, pp. 192-253, December 1955; Vol. 4, No. 2, pp. 86-119, June 1956

E. DESCRIPTIONS OF SPATIAL STRUCTURE

1. INTRODUCTION

Previous sections have discussed the representation of functions where time was implicitly or explicitly indicated as the independent variable. A number of descriptions and relationships were outlined. Their use was indicated to be jointly dependent on the type of function, and on the use which was to be made of the descriptions. Conditions under which deterministic and statistical analyses were preferred or permissible were given. The existence of boundaries in time, or in the conjugate domain, frequency, was seen to play an important part in determining the type of description which could be used for a particular problem. Bounds imposed jointly on the conjugate domains involved other important classes of descriptions. Additional representatives were discussed which were useful when a multiplicity of functions were involved.

Many of these descriptions and the problems associated with their selection and use also occur in spatial problems — that is, where the independent variables are spatial. It is not intended to develop the correspondence of spatial descriptions completely — and only a few elementary aspects will be reviewed in this section. Other relationships will be developed in later sections. Although the exact details of these relationships are of importance, of comparable significance is the understanding of the basic philosophy associated with concepts such as the transformation of spatial variables, and spatial sampling.

Just as time and frequency were seen to provide equally useful methods for description, both space and space frequency are employed in representing spatial structure. A spatial distribution may be expressed mathematically as a function of intensity and of linear position along a line or in a plane. It may also be expressed mathematically as a function of inverse space, that is, in terms of spatial frequencies. The relative utility of the two domains is based on essentially the same factors which make Fourier transformation of value for temporal functions. That is, it may often be of value in order to improve visualization, or to facilitate computation or measurement to work in the space-frequency domain rather than directly in the space domain. Similarly, correlation analyses and sampling methods may be applied to simplify representations of spatial structural detail.

2. FOURIER ANALYSIS

AMPLITUDE DISTRIBUTION; RADIATION PATTERN

A well-known example of the application of Fourier analysis to spatial problems involves the relationship between the amplitude distribution along an aperture and the angular distribution of energy. For antennas, the distribution along the aperture is given by the component of the excitation tangential to the aperture plane that produces or maintains a radiation field at an arbitrary point in space. Any amplitude or aperture distribution $F(x)$ may be represented by the following expression:

$$F(x) = A \cos \theta \, e^{-jk \sin \theta x} \qquad \text{(E-1)}$$

This represents waves traveling over the aperture and having different propagation coefficients along the x-axis. The equation above represents a wave with a propagation coefficient $k \sin \theta$ which produces in the medium (propagation coefficient k) a plane wave in a direction making an angle θ with the normal to the aperture plane, A is a complex number whose modulus and argument determine the amplitude and phase of the plane wave at the aperture, and k

represents the increase of phase difference per unit distance in the direction of propagation. Each wave of the Fourier decomposition has its own amplitude and phase, which in general vary with θ, forming an angular spectrum of plane waves.

The concept of a polar diagram implies an aperture of finite dimensions and the evaluation of the field at a point whose distance from the aperture is large compared to the extent of the aperture and the wavelength. Under these conditions the angular spectrum is called a polar diagram. It is important to note that the angular spectrum associated with an aperture distribution gives the polar diagram if this concept is applicable but retains a useful meaning even when it is not possible to use a polar diagram as a method of representation.

For an aperture of width "a" having a real or complex amplitude distribution $F(x)$, the radiation pattern $G(s)$ is given by

$$G(s) = (1 + \cos\theta) \int_{-a/2}^{a/2} F(x)\, e^{j 2\pi x \frac{\sin\theta}{\lambda}}\, dx \tag{E-2}$$

where

θ = the polar angle measured from the normal to the aperture

x = the distance along a plane parallel to the aperture

λ = the wave length

$s = -\sin\theta/\lambda$.

It is assumed that the phase velocity at the aperture is equal to the velocity of propagation. If the beam is sufficiently narrow, the slowly varying factor $1 + \cos\theta$ can be omitted and diffraction theory used to predict the radiation pattern that will be obtained with a given aperture excitation and aperture width. Fourier analysis of the aperture distribution gives the position and strength of the component beams. The line spectrum is an angular one, $\sin\theta/\lambda$ replacing the frequency variable of ordinary harmonic analysis. The lines represent plane waves which would be produced if the distribution of the field over the aperture plane was periodic extending to infinity. When the distribution is confined to a single period the effect is approximately that of forcing plane waves through a finite aperture. The angular spectrum then becomes continuous, and each plane wave is replaced by a diffraction pattern in the form of a main beam and sidelobes. Thus, a Fourier series representation of the field is transformed into a Fourier integral, the aperture distribution and radiation pattern together comprising a pair of Fourier transforms. If the aperture is fed in-phase, a one-dimensional aperture distribution (or amplitude distribution) and radiation pattern (or angular spectrum) may be represented as:

$$G(s) = \int_{-\infty}^{\infty} F(x)\, e^{-j 2\pi x s}\, dx \quad \text{(radiation pattern)} \tag{E-3}$$

$$F(x) = \int_{-\infty}^{\infty} G(s)\, e^{j 2\pi x s}\, ds \quad \text{(aperture distribution).} \tag{E-4}$$

If the amplitude distribution is symmetrical about the center of the aperture, $F(x)$ is even and the pattern is given by the Fourier cosine transform of $F(x)$, designated as $G_c(s)$, then (E-3) and (E-4) becomes

$$G_c(s) = \int_{-\infty}^{\infty} F(x) \cos(2\pi x s)\, dx \tag{E-5}$$

$$F(x) = \int_{-\infty}^{\infty} G_c(s) \cos(2\pi x s)\, ds. \tag{E-6}$$

If $F(x)$ is odd, the excitation of one half is of opposite sign to that of the other half, and the radiation pattern will be given by the Fourier sine transform of $F(x)$, i.e., $G_s(s)$:

$$G_s(s) = \int_{-\infty}^{\infty} F(x) \sin(2\pi xs)\, dx \tag{E-7}$$

$$F(x) = \int_{-\infty}^{\infty} G_s(s) \sin(2\pi xs)\, ds\,. \tag{E-8}$$

Fourier sine and cosine transform pairs are illustrated in Figures E-1, E-2, and E-3 for several types of amplitude distributions. The antiphased apertures are used in "Monopulse" application as will be indicated in later sections.

A given radiation pattern may be resolved into the sum of two or more radiation patterns, each supplying its own aperture distribution. The resulting aperture distribution will then be the vector sum of the component distributions. An elementary and useful pattern is the "sin x-over-x" pattern which results from a constant amplitude, in-phase distribution across the aperture. This is shown in Figure E-1. The zeros are equally spaced, except for those on either side of the main beam, which occupies two "spaces." The width of one "space" is the reciprocal of the aperture width, since the beamwidth is inversely proportional to the width of the aperture.

Two other properties of the Fourier transform which are of importance in antenna problems:

(1) Delay Linear Added Phase

$$F(x-a) \rightleftharpoons G(s)\, e^{-j2\pi as}$$

Similar to the Fourier property in the time-frequency domain, a translation in the aperture distribution affects only the phase of the radiation pattern.

(2) Complex Modulation Shift of Spectrum

$$F(x)\, e^{j2\pi\gamma x} \rightleftharpoons G(s-\gamma)\,.$$

Multiplying an aperture distribution by exponent $(j2\pi\gamma x)$ "delays" or displaces the radiation pattern by an amount θ_o where $\gamma = -\sin\theta_o/\lambda$. This results in having the axis of the major lobe of the radiation pattern at an angle of θ_o with respect to the normal to the aperture. If the total phase variation across the aperture amounts to $2n\pi$, n being an integer, the pattern is displaced n spaces.

Any number of main beams of the "sin x-over-x" patterns, if each is separated from the others by a whole number of spaces, may be summed without interferring with each other in their principal direction of radiation. Hence, all zeros may be made to disappear by placing beams adjacent to each other at intervals of one space. Therefore, for an aperture of n wavelengths, a polar diagram may be constructed so as to have any chosen value in each of $2n + 1$ different directions (equally spaced) in front of the aperture.

APERTURE - BEAMWIDTH PRODUCT

A relationship between the aperture distribution width and the width of the angular spectrum of the radiation field is,

$$\Delta s \times \Delta a = 1 \tag{E-9}$$

where Δs = the equivalent angular spectrum width

Δa = the equivalent aperture distribution width.

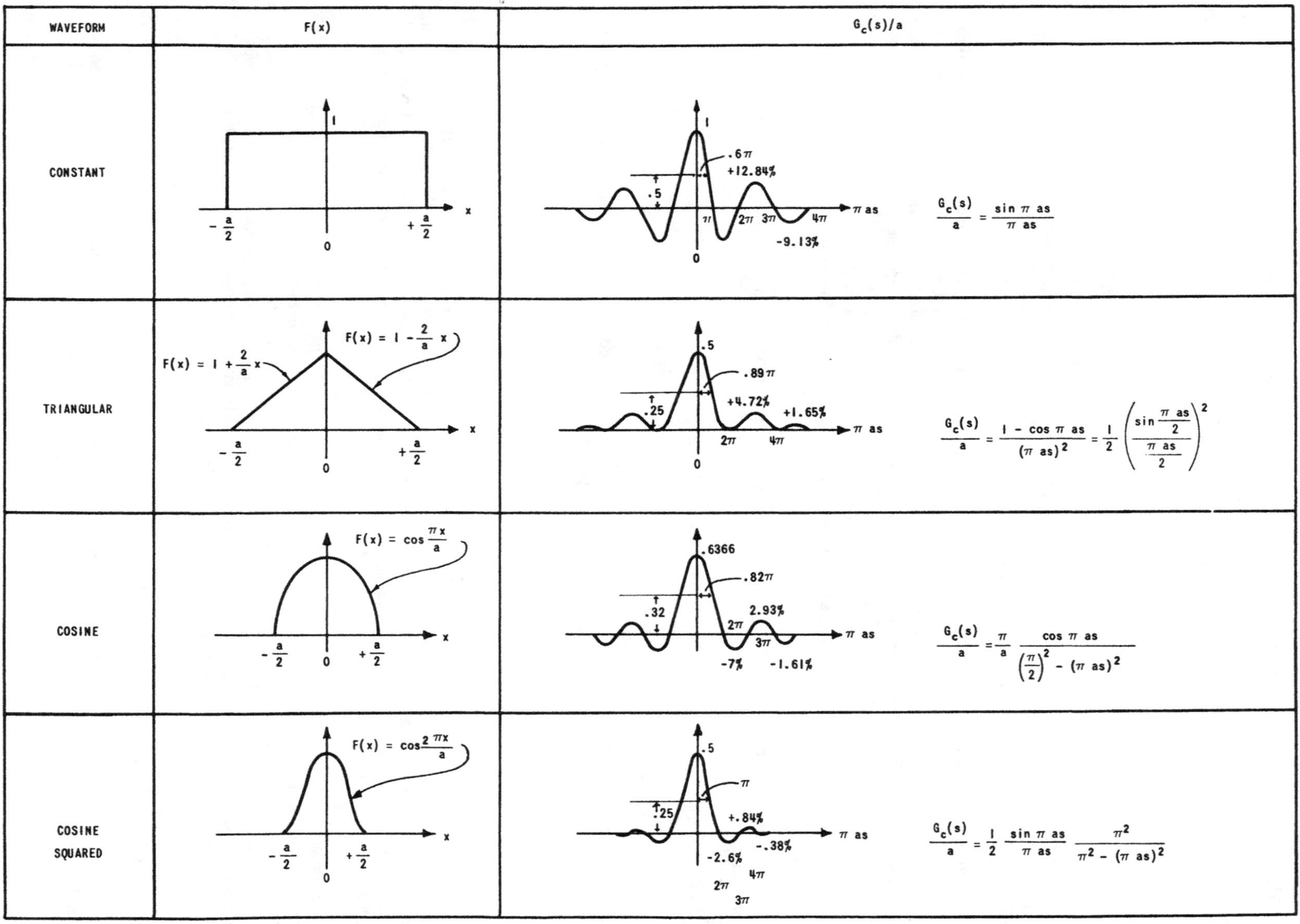

Figure E-1 - Fourier cosine transforms

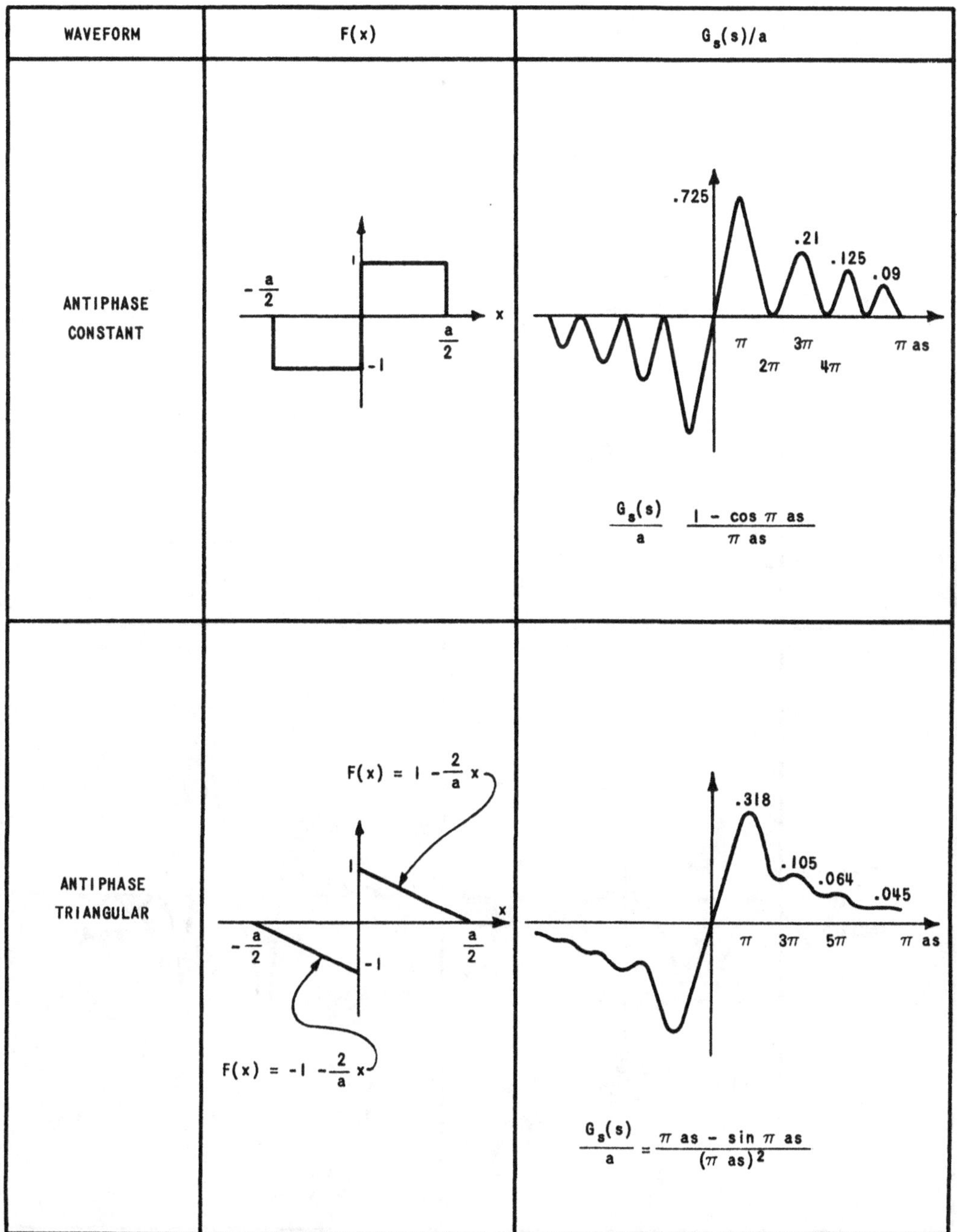

Figure E-2 - Fourier sine transforms

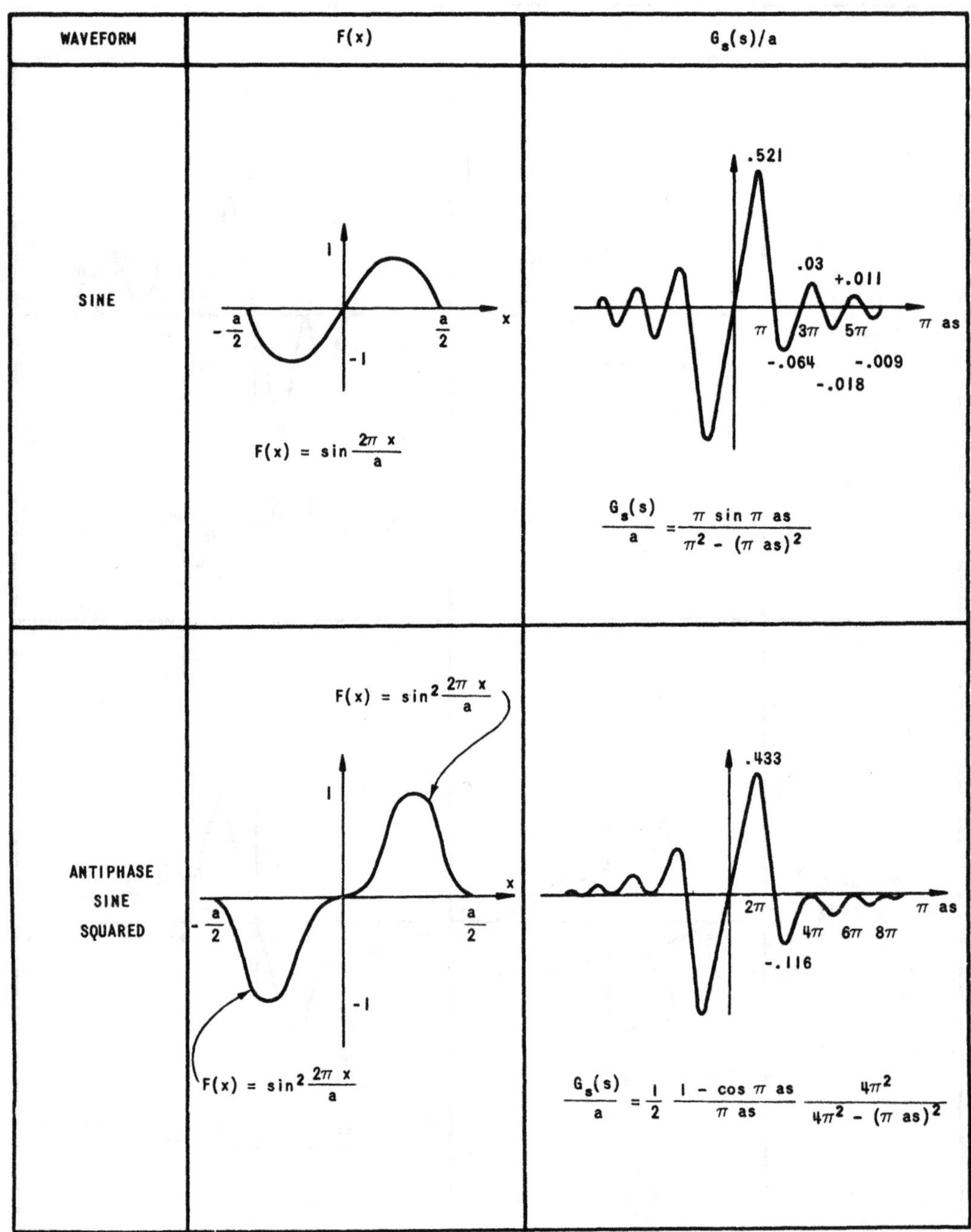

Figure E-3 - Fourier sine transforms

This is valid provided that $G^2(0) \neq 0$, where $G(0) = G(s)|_{s=0}$ and is the angular spectrum of plane waves radiated from the aperture at broadside.

Thus, an aperture-beamwidth reciprocal relationship exists for a spatial distribution, analogous to that for the time-frequency domain. It has been previously shown that the radiation pattern due to a constant, in-phase, amplitude (unity) distribution over an aperture of width "a" is

$$G(s) = a\,\frac{\sin \pi as}{\pi as} \tag{E-10}$$

Comparing this with the spectral envelope of a rectangular pulse of unit amplitude and pulse width T, namely,

$$G(f) = T\,\frac{\sin \pi Tf}{\pi Tf} \tag{E-11}$$

shows that the aperture width "a" corresponds to the pulse width T and the direction parameter s corresponds to frequency f. To complete the analogy, replace the time variable t applying to the pulse by the distance x along the aperture. If the Fourier transform relationships between waveform and spectrum in the time domain are:

$$G(f) = \int_{-\infty}^{\infty} F(t)\exp(-j2\pi ft)\,dt \quad \text{(frequency spectrum)} \tag{E-12}$$

$$F(t) = \int_{-\infty}^{\infty} G(f)\exp(+j2\pi ft)\,df \quad \text{(time function)} \tag{E-13}$$

then using the above analogies, the corresponding Fourier transforms in the space domain are:

$$G(s) = \int_{-\infty}^{\infty} F(x)\exp(-j2\pi xs)\,dx \quad \text{(radiation pattern)} \tag{E-14}$$

$$F(x) = \int_{-\infty}^{\infty} G(s)\exp(+j2\pi xs)\,ds \quad \text{(aperture distribution).} \tag{E-15}$$

which are identical to (E-3) and (E-4).

To determine the effect of varying the aperture width, consider the radiation pattern given by (E-10). If the aperture width is increased by a factor "m", the new pattern is

$$G(s) = ma\,\frac{\sin \pi mas}{\pi mas}\ . \tag{E-16}$$

Therefore, increasing the aperture m times has the following results:

1. the field strength increases m times at broadside

2. the beamwidth of the angular spectrum is reduced by 1/m

3. the beamwidth in the polar diagram decreases but not linearly since the half-power beamwidth depends on the arc-sine; the reduction will be linear for narrow beams where $\theta \approx \sin\theta$ ($\theta < 15°$). The inverse will be obtained if the aperture width is decreased m times. These results are illustrated in Figure E-4.

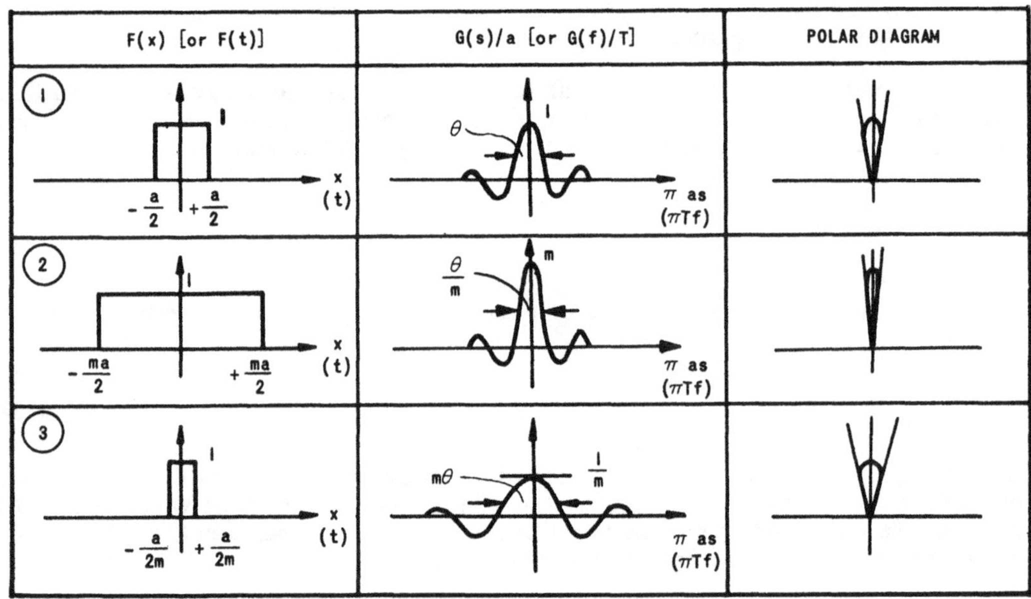

Figure E-4 - Reciprocal relationship between aperture and beamwidth

TOTAL RADIATION FIELD

The Fourier methods discussed are rigorously applicable only in the far field, that is, where the energy can be considered to be propagated by plane waves. Calculations of the complete field can become extremely complex with exact solutions being obtainable only for relatively simple configurations. The increase in the physical size of many radar and sonar arrays has made it a practical necessity to consider approximate methods of describing and relating near- and far-field radiation.

The total radiation field may be arbitrarily divided into three "regions," the Fraunhofer region or far field, the Fresnel region and the near field. These regions are distinguished by the nature of the approximations made for establishing the functional dependence of the field contribution of an aperture element on the separation of the element to a point in space.

The far field is a region in which power decreases inversely as the second power of the distance to the center of the aperture. The approximation made is that the field contribution due to an aperture element consists of a constant amplitude factor and a linear phase factor. The radiation pattern in this region is then the Fourier transform of the aperture distribution as was indicated earlier in the section.

Within the Fresnel region an element contribution may be approximated by a constant amplitude and a linear and quadratic factor. The Fresnel region extends, with respect to the center of the aperture, from a distance of several times its diameter D to D^2/λ, where λ is the wavelength of the propagating wave.

The near field extends from the Fresnel region to the surface of the aperture. It is characterized by extreme amplitude and phase variations, and has been designated the "induction" field in electromagnetics. For acoustical sources having dimensions small compared to a wavelength the near field would extend to a distance of one-quarter wavelength beyond the source. The distance from the source at which the transition from the Fresnel to the Fraunhofer region occurs is, of course, not sharply defined physically, and is dependent on the shape of the source and the amplitude distribution. In the Fraunhofer region, the power on axis is inversely proportional to the square of the distance, whereas in the Fresnel region, a series of maxima and minima will be found on the axis normal to the plane of the source. More detailed discussions of the significance of these regions in information processing will be given in later sections.

SPATIAL SAMPLING

Sampling theorems which were previously discussed were related to Fourier analysis. One of the theorems indicated that a function restricted to an upper frequency of W could be completely described by giving its values at a series of points spaced $1/2W$ apart.

A version of spatial sampling may be stated in the following form:

A discrete source confined within $x = \pm X_c$, $y = \pm Y_c$ is completely specified by measuring the coherence of its field at discrete invervals at $1/2X_c$ and $1/2\,Y_c$, where X_c and Y_c are the widths of the aperture in wavelengths along planes parallel to the x and y directions respectively. This results in a two-dimensional radiation pattern. (By a discrete source is meant one having finite dimensions and a finite aperture distribution.)

The above theorem gives the interval for which independent measurements are to be made. Measurements could be made at a finer interval. However, the measurements at finer intervals would be deducible from measurements at the greatest interval compatible with the theorem, and would not be fully independent. As a result, the structural information of a field, contributed by a source, can be thought of being spread out over the medium and having a certain density, there being one independent datum per rectangular cell having the dimensions given above.

The space-sampling intervals for two types of distributions are given below:

Constant distribution, in-phase

$$\theta_{1/2} = \frac{69}{a/\lambda} \text{ degrees } = \text{ half-power beamwidth}$$

$$\theta_s = .83\theta_{1/2} \text{ degrees } = \text{ space-sampling interval.}$$

Cosine distribution, in-phase

$$\theta_{1/2} = \frac{94}{a/\lambda} \text{ degrees } = \text{ half-power beamwidth}$$

$$\theta_s = .61\theta_{1/2} \text{ degrees } = \text{ space-sampling interval.}$$

Thus, it is seen that the space-sampling interval is related to the beamwidth of the radiation pattern.

In the time-frequency relationships, it was indicated that when a sampled waveform is passed through a low-pass filter which transmits only the spectrum of the original signal, each individual delta-function (a sample) having an infinitely wide spectrum was reduced to a $\sin x/x$ function, whose rectangular spectrum fits the filter pass band. Therefore, the original signal is the superposition of a series of $\sin x/x$ functions spaced according to the sampling theorem weighted by the filtered waveform. In space, a finite aperture acts like a low-pass filter and results in a linear distortion of the radiation pattern as illustrated in Figure E-5. Each sample of the radiation pattern when passed through a finite aperture will result in a $\sin x/x$ shaped beam that corresponds to a direction in the radiation field. Figure E-6 illustrates how a set of sample beams can synthesize the radiation pattern.

AUTOCORRELATION FUNCTION

The autocorrelation function $\psi(x)$ of an amplitude distribution $F(x)$ is given as

$$\psi(x') = \frac{1}{L}\int_{o}^{L} F(x)\,F(x+x')\,dx \qquad \text{(E-17)}$$

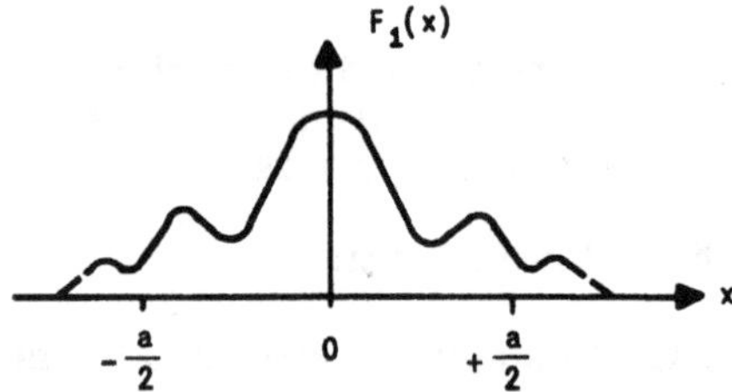

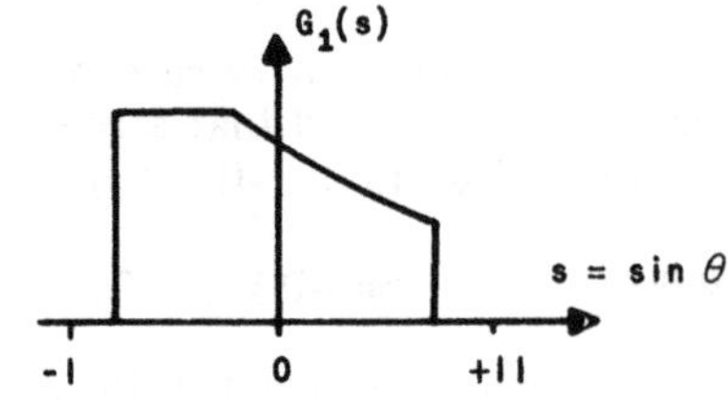

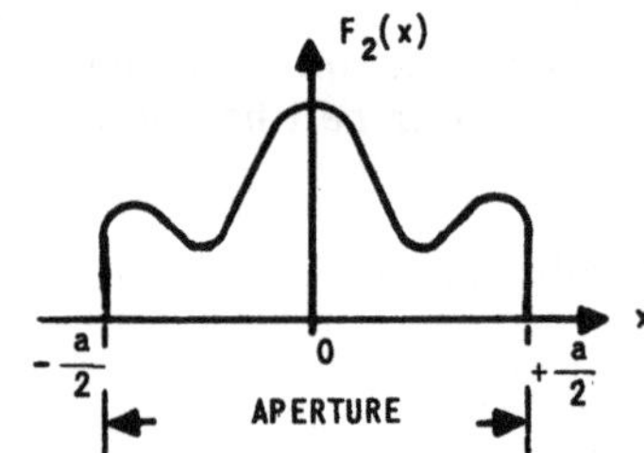

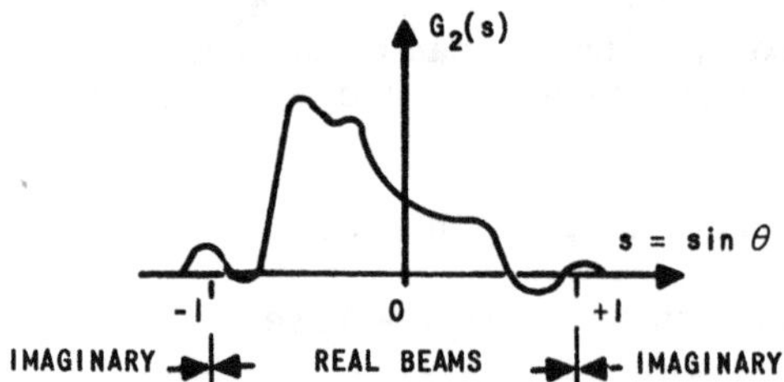

Figure E-5 - The radiation pattern $G_1(s)$ and its aperture distribution $F_1(x)$ are shown together with the truncated aperture distribution $F_2(x)$ and its corresponding radiation pattern $G_2(s)$

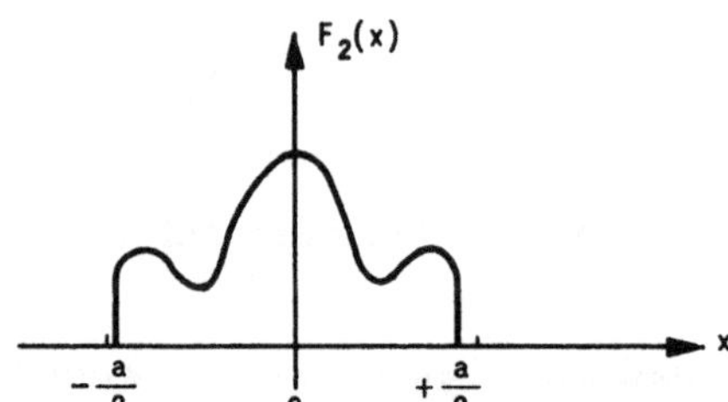

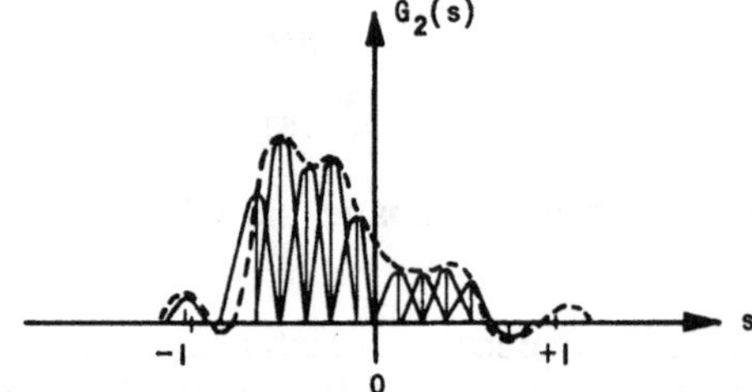

Figure E-6 - A set of sample beams which synthesize the realizable beam pattern $G_2(s)$. The sidelobes of the sample beams are omitted.

where L is the length of the line. Similar to the time domain, the autocorrelation function at the origin is the mean-square-value of the amplitude distribution. If x' is larger than the distance over which the value of one point in $F(x)$ has an influence on the other, then $\psi(x')$ approaches the square of the average value of $F(x)$. In two dimensions, autocorrelation functions may be used as measures of the linear coherence of two-dimensional amplitude distribution functions and are expressed as

$$\psi(x',y') = \frac{1}{A} \iint_A F(x,y)\ F(x+x',\ y+y')\ dx\ dy \qquad \text{(E-18)}$$

where A is the area of the field under consideration.

Correlation analysis may be used to describe the statistical properties of the fluctuations of a wave field. Previously it was shown that the correlation function can be used to measure the "time constant" of a waveform. Similarly, a "correlation distance" in the space domain may be used to describe the distance at which the statistical dependence between fluctuations vanishes. The amplitude and phase variations in a radiation field are determined to a large extent by the type of background ensembles existing in the field. These can be classified in many ways. For example, an ensemble of backgrounds may be stationary - the properties do not vary with the direction of view, ergodic - the statistical properties of any background are the same as those of any other of the ensemble, or Gaussian - the properties are analogous to those of electrical noise generated in radiation detectors.

The spatial autocorrelation function and its Fourier transform, the power pattern, are very useful for decribing Gaussian background ensembles and can be used to "optimize" linear spatial systems in the least mean squares sense. For non-Gaussian ensembles, correlation analysis cannot completely describe the field, Also, it cannot be used to analyze the performance of non-linear spatial systems. Non-Gaussianity or non-linearity infers statistical descriptions for complete characterization.

REFERENCES

E-1. J. W. Horton, "Fundamentals of Sonar," United States Naval Institute, 1957

E-2. J. F. Ramsey, "Fourier Transforms in Aerial Theory," Marconi Review, No. 83, p. 139; No. 84, p. 17; No. 85, p. 41; No. 86, p. 81; No. 87, p. 157; No. 89, p. 45; 1946-1948

E-3. P. M. Woodward, "A Method of Calculating the Field Over a Plane Aperture Required to Produce a Given Polar Diagram," J.I.E.E., Vol. 93, Part III A, 1956, pp. 1554-1558

E-4. H. G. Booker and P. C. Clemmow, "The Concepts of an Angular Spectrum of Plane Waves, and its Relation to that of Polar Diagram and Aperture Distribution," Proc. I.E.E., Vol. 97, Part 3, 1950, pp. 11-17

E-5. R. C. Hansen and L. L. Bailin, "A New Method of Near Field Analysis," I.R.E. Trans., Vol. AP-7, Dec. 1959, Special Supplement, pp. S458-S467

E-6. G. T. DiFrancia, "Resolving Power and Information," J. Opt. Soc. Am., Vol. 45, No. 7, July 1955, pp. 497-501

E-7. R. N. Bracewell, "Radio Interferometry of Discrete Sources," I.R.E. Proc., Jan. 1958

E-8. H. P. Raabe, "Antenna Pattern Synthesis of the Most Truthful Approximation," I.R.E. Wescon Convention Record, Part I, 1958, pp. 178-183

E-9. H. L. Knudson, "Shannon's Time-Division Formula and Superamplification in Antennas," Teknisk Tidskrift 2.12.52, pp. 1023-1030

E-10. G. T. DiFrancia, "Directivity, Super-Gain and Information," Electromagnetic Wave Theory Symposium, July 1956

E-11. L. Clayton, Jr., and J. S. Hollis, "Calculation of Microwave Antenna Radiation Systems By the Fourier Integral Method," The Microwave Journal, Sept. 1960, pp. 59-65

E-12. P. M. Woodward and J. D. Lawson, "The Theoretical Precision With Which an Arbitrary Radiation Pattern May Be Obtained From a Source of Finite Size," Proc. I.R.E., Part III, Vol. 95, 1948, pp. 363-370

E-13. D. Z. Robinson, "Methods of Background Description and Their Utility," I.R.E. Proc., Sept. 1959, pp. 1554-1561

E-14. J. B. Smyth, "Space Analysis of Radio Signals," Journal of Research of the NBS, Vol. 650, No. 3, May-June 1961

E-15. D. R. Rhodes, "Introduction to Mono-Pulse," McGraw Hill Book Company, Inc., 1959

E-16. L. Brillouin, "Science and Information Theory," Academic Press, pp. 105-112, 1956

E-17. F. Zucker, "Summary Comments," Proceedings of Symposium on Communication Theory and Antenna Design, AFCRC-TR-57-105

F. ORTHOGONALITY AND INTEGRAL TRANSFORMS

1. INTRODUCTION

The equations that describe physical phenomena ordinarily depend on both time and space coordinates. By introducing parameters K_n called "proper values" the equations may be transformed to those depending on the space coordinates only. These values, which are usually infinite in number, are determined from conditions which must be satisfied at certain physical boundaries.

The particular form of these equations depends upon the system of coordinates used, and the choice of coordinates in turn depends upon the geometry of the physical system to which the equations apply. The particular types of functions which satisfy the equations are known by names which refer to the particular geometry of the physical system. Examples of these are (a) cylindrical functions, of which the Bessel functions are of the "first kind" and (b) spherical harmonics, also known as Legendre' Polynomials.

The functions are referred to as the "proper functions" or "eigenfunctions" pertaining to the particular physical system under consideration, the simplest of them being the trigonometric functions which are the proper functions for systems having a rectangular geometry. If a function $f(x,y,z)$ is operated upon by a linear operator $L\,[f(x,y,z)]$, such that $L\,[f(x,y,z)] = (\text{constant})\ f(x,y,z)$, then the function is called an eigenfunction of the transformation, and the constant the corresponding eigenvalue or proper value, K_n. These functions, in terms of the proper values K_n, form a set or system. Due to the linearity of the equations, a complete solution may be derived by a linear superposition of a set of these proper functions with different parameter values and arbitrary coefficients. Thus, if $\phi_n(x,y,z)$ represents a proper function for the parameter n, the solution $f(x,y,z)$ will have the form

$$f(x,y,z) = a_1\phi_1 + a_2\phi_2 + a_3\phi_3 + \cdots + a_n\phi_n \tag{F-1}$$

which in general is an infinite series. The coefficients a_n are regarded as constants of integration which give the solution (F-1) the necessary flexibility of meeting certain boundary conditions set by the physical problem.

In one dimension, (F-1) takes the form

$$f(x) = \sum_{n=0}^{\infty} a_n\phi_n(x)\,. \tag{F-2}$$

The problem now is to expand $f(x)$, an arbitrary function, in a series of weighted elementary proper functions in such a way that the resulting series converge. The solution to this problem is usually quite complicated unless the system of proper functions or a derived system formed from linear combinations of these functions satisfies what are known as the conditions of orthogonality. Before these conditions are given, the origin of the word orthogonality will be discussed.

2. CONDITIONS OF ORTHOGONALITY

The word orthogonality comes originally from vector analysis where two vectors $A(x,y,z)$ and $B(x,y,z)$ are said to be orthogonal if their dot product equals zero, i.e.,

$$A \cdot B = A_x B_x + A_y B_y + A_z B_z = 0\,. \tag{F-3}$$

Similarly, vectors in n dimensions having components A_i, B_i, $(i = 1,2,3, \ldots n)$ are said to be orthogonal when

$$\sum_{i=1}^{n} A_i B_i = 0. \qquad \text{(F-4)}$$

If a vector space has an infinite number of dimensions the components A_i and B_i become continuously distributed and i is no longer a denumerable index but a continuous variable (x). If (x) is confined to the region $0 \le x \le \ell$, the scalar product (F-4) becomes

$$\int_0^{\ell} A(x)\, B(x)\, dx = 0. \qquad \text{(F-5)}$$

In this case the functions $A(x)$ and $B(x)$ are said to be orthogonal. The concept of orthogonality is indefinite unless reference is made to specific range of integration which in the present case is from 0 to ℓ.

In general, the conditions of orthogonality for the one-dimensional case are given as

$$\int_a^b W(x)\, \phi_m(x)\, \phi_n(x)\, dx = \begin{cases} r_n & \text{for } m = n \\ 0 & \text{for } m \neq n \end{cases} \qquad \text{(F-6)}$$

where $W(x)$ is a fixed function of the independent variable which is usually taken equal to unity when representing signals, and a and b are the finite limits of the region over which the function $f(x)$ is specified. To obtain the coefficients, multiply (F-2) by $W(x)\,\phi_n(x)$ and integrate from $x = a$ to $x = b$,

$$\int_a^b W(x)\, f(x)\, \phi_m(x)\, dx = \sum_{n=0}^{\infty} a_n \int_a^b W(x)\, \phi_m(x)\, \phi_n(x)\, dx\,. \qquad \text{(F-7)}$$

Since the ϕ's are orthogonal, from (F-6), all terms on the right of (F-7) vanish except one so that

$$\int_a^b W(x)\, f(x)\, \phi_m(x)\, dx = a_n \int_a^b W(x) \left[\phi_m(x)\right]^2 dx. \qquad \text{(F-8)}$$

Solving for a_n,

$$a_n = \frac{\int_a^b W(x)\, f(x)\, \phi_m(x)\, dx}{\int_a^b W(x) \left[\phi_m(x)\right]^2 dx}. \qquad \text{(F-9)}$$

Since an orthogonal function $\phi_m(x)$ may be multiplied by an arbitrary constant, the quantity r_n can be made equal to unity. The resulting functions $\phi_n(x)$ are then referred to as a normalized set, the denominator of (F-9) becomes unity and the coefficients for (F-2) are

$$a_n = \int_a^b W(x)\, f(x)\, \phi_n(x)\, dx \qquad \text{(F-10)}$$

which represents the desired solution to the problem stated earlier.

3. INTEGRAL SQUARE ERROR

The question now arises under what circumstances it is possible to express an arbitrary signal $f(x)$ by an infinite sequence of orthogonal functions $\phi_n(x)$,

$$f(x) = \sum_{-\infty}^{\infty} a_n \phi_n(x) . \tag{F-11}$$

In order to use a representation such as (F-11), it is necessary to either restrict $f(x)$ or settle for something less than an exact identity in the representation.

A satisfactory way of specifying the near equality of $f(x)$ and the sum is through an integral of the magnitude of the difference squared:

$$\int_a^b W(x) \left| f(x) - \sum_{-N}^{N} a_n \phi_n(x) \right|^2 dx = \epsilon. \tag{F-12}$$

If the a_n can be so chosen that the integral (F-12) vanishes for a given function, then the representation (F-11) is said to be complete.

The use of a criterion of this type to evaluate the effectiveness of a representation is arbitrary. The odd powers cannot be used for then ϵ has no minimum and although any even power could serve the purpose, the evaluation of the coefficients would become very complicated. Too, a power law greater than the square law would tend to suppress very large errors at the expense of smaller errors to a greater extent than the square law. The latter treats all errors more equally and, as will soon be seen, lends itself to the concept of orthogonality.

To find the coefficients which minimize ϵ, expand (F-12), that is,

$$\epsilon = \int_a^b W(x) \left| f(x) \right|^2 dx - 2 \sum_{-N}^{N} a_n \int_a^b W(x)\ f(x)\ \phi_n(x)\ dx + \sum_{m,n=-N}^{N} a_m a_n \int_a^b W(x)\ \phi_m(x)\ \phi_n(x)\ dx. \tag{F-13}$$

If $\phi_n(x)$ is orthogonal and normalized so that

$$\begin{aligned} \int_a^b W(x)\ \phi_m(x)\ \phi_n(x)\ dx &= 0 \qquad m \neq n \\ &= 1 \qquad m = n \end{aligned} \tag{F-14}$$

then (F-13) becomes

$$\epsilon = \int_a^b W(x) \left| f(x) \right|^2 dx - 2 \sum_{-N}^{N} a_n \int_a^b W(x)\ f(x) \cdot \phi_n(x)\ dx + \sum_{-N}^{N} \left| a_n \right|^2 \tag{F-15}$$

which, from (F-10), reduces to

$$\epsilon = \int_a^b W(x) \left| f(x) \right|^2 dx - \sum_{-N}^{N} \left| a_n \right|^2 \tag{F-16}$$

Thus, if the ϕ_n are orthonormal functions the error becomes a minimum and as N becomes infinite, (F-16) results in Parseval's theorem. In general, the theorem states that if an arbitrary function is expressed as an infinite weighted sum of orthogonal functions, then the "energy" of the function is equal to the sum of the "energy" of each of the orthogonal components.

When (F-11) is not exactly an equality, (F-16) suggests that the orthogonal function method of finding the coefficients leads to the minimum value for the integral of the square of the discrepancy.

4. INTEGRAL TRANSFORMS

If the function $f(t)$ is defined by an ordinary or partial difference, differential, or integral equation and certain boundary conditions, it is found simpler in certain circumstances to translate the boundary value problem for $f(t)$ into one for the function,

$$F(s) = \int_{t_1}^{t_2} f(t)\, K(s,t)\, dt \qquad \text{(F-17)}$$

where $F(s)$ is called the integral transform of $f(t)$, $K(s,t)$ the kernel function, and s the image variable. From (F-17), an integral transformation of a function $f(t)$ is obtained by:

(1) Multiplying $f(t)$ by a function of two variables.

(2) Integrating over a definite range of the original independent variable so that the transformation is a function of the image variable only.

The integral transformation is orthogonal for it satisfies the conditions of orthogonality and consequently may be evaluated by the integral-square-error criterion for completeness. It provides a one-to-one transformation with the function being transformed.

The kernel function determines the type of integral transformation made and the applicability of the transform to particular systems. There is a great variety of functions which may be used as kernels. Three general categories are:

(1) Product kernels — the original and image variables occur as a product st.

(2) Sum or difference kernels — the original and image variables occur as a sum or difference $s \pm t$.

(3) Types in which the original and image variables do not occur in a combination that can be replaced by a single variable.

In a transformation, the independent physical variable (such as time t, distance x, temperature T, etc.) is replaced by an abstract mathematical variable called the image variable (usually represented as s, p, or $j\omega$) and the dependent physical variable is then replaced by an abstract function called the transform. Physical significance has been attached to these abstract variables, the extent of which depends on their utility.

In order to obtain the unknown function from its transform it is necessary to invert or solve the integral equation (F-17). The general method of inversion is by an inversion integral of the form

$$f(t) = \int_{s_1}^{s_2} k(t,s)\, F(s)\, ds. \qquad \text{(F-18)}$$

The inversion integral must have this general form since:

1. It must contain the transform of the unknown function, $F(s)$.
2. It must have a kernel function of s and t in order that the integral be a function of t.
3. It must have definite limits so that it is not a function of s.

The transform integral and the inversion integral together constitute an integral equation pair.

One procedure for producing integral equation pairs is through the use of spectral theory. The technique involved is that the function $f(t)$ is expanded in terms of a set of discrete functions $k_n(t)$ that possess an orthogonality property, where n has a different integer value for each member of the set. Such sets of functions are called eigenfunctions or spectral functions. The expansion is

$$f(t) = \sum_n k_n(t)\, f_n \qquad \text{(F-19)}$$

where f_n, the coefficient of $k_n(t)$ in the expansion, is the amount of each spectral function which must be present in order that the superposition add up to the given function, $f(t)$. The set of coefficients f_n is called the spectrum of $f(t)$ with respect to the set of spectral functions in terms of which $f(t)$ is resolved. Since the set of spectral functions is discrete, the spectrum is called a discrete spectrum.

When the number of terms in the spectrum required to give a good approximation of the function being resolved is small, the summing process can be performed. However, if the number of terms needed for a good approximation are large, it is convenient to convert the sum to an integral. Therefore, the above becomes

$$f(t) = \int k_n(t)\, df_n \qquad \text{(F-20)}$$

As the number of spectral terms required for an adequate representation increases indefinitely they form a denser set of spectral lines and in the limit become a continuous spectrum. In this process, the discrete variable n may be replaced by a continuous variable s and the discrete set of spectral functions $k_n(t)$ may be represented as a continuous set of spectral functions $k(t,s)$. Therefore, (F-20) becomes:

$$f(t) = \int k(t,s)\, df_s. \qquad \text{(F-21)}$$

But df_s can be represented as a derivative in spectral space:

$$df_s = \frac{df_s}{ds}\, ds \qquad \text{(F-22)}$$

where df_s/ds is the spectral intensity or amplitude density in spectral space and is called the continuous spectrum of $f(t)$ with respect to the spectral or kernel function $k(t,s)$. The spectral intensity of a function is represented as $F(s)$ so that (F-21) can be put in conventional form as

$$f(t) = \int_{s_1}^{s_2} k(t,s)\, F(s)\, ds \qquad \text{(F-18)}$$

which is the inversion integral.

In order to obtain an expansion of $f(t)$ in terms of a discrete or continuous set of spectral functions, it is necessary to determine the spectral amplitude f_n or the spectral intensity $F(s)$. For the coefficients f_n to be obtained easily, the set of spectral functions must have the property of orthogonality. The concept of a transformation, therefore, hinges on expressing an arbitrary function with a series of suitable functions which possess the property of orthogonality.

A complete transformation involves three operations. These are:

(1) Make transformation
(2) Obtain solution in transformed domain
(3) Perform inverse transformation.

The first operation characterizes the transformation and the problem to which it is being applied. The transformation made must serve the purpose of matching the source of the problem to its application. For instance, when dealing with numbers, "number" transforms such as logarithms are used, whereas with functions, "function" transforms such as the Fourier, Laplace, Mellin, Hankel, and Z transforms are employed. The type of kernel the transform has determines the function or system to which it may be applied. The second and third operations provide the reduction of mathematical complexity or improved visualization of the problem.

Table F-1 lists differential, integral, and difference equations which are of importance in many physical applications and which relate the response of a linear element to its excitation. The type of transform best suited for each equation and the results obtained through its use is also given. It is seen that the Laplace transform is effective in reducing an nth order differential equation with mth degree polynomial coefficients to one of reduced order in the transform domain where it is assumed that n is greater than m. The Laplace transform, however, is usually used to solve a linear constant coefficient differential equation or integral equation of the Volterra type. Both become algebraic in the transformation. It is also used to aid solving linear time-variant systems where the time-varying parameter is a function of the first power of (t). In this case a first-order differential equation in the transformed domain is obtained. In general, the Laplace transform enables the steady state solution, the transient solution and all initial conditions to be treated in one single operation. This is in contrast to the classical method of solving for the source-free and forced solutions individually, summing them to give the general solution, and evaluating the constants of integration from the initial conditions.

The Mellin and Hankel transforms lend themselves to solving problems concerning time-varying linear systems. In particular, the Hankel transform is used to solve Bessel-type differential equations since its kernel is a Bessel-function. The Mellin transform is applied successfully to systems characterized by an Euler-Cauchy differential equation or Fredholm integral equation reducing both to algebraic form in the conjugate domain. A Hankel transform pair is symmetric and thus only one table is necessary for the direct and inverse transformations. That is, the same column can represent either the function or its transform. This is not true of the Laplace or Mellin transforms since the transform variable is complex and the inverse transform is obtained by performing a complex integration.

The Z transform is used to solve linear difference equations. The difference equation applies to a discrete signal system and is analogous to differential and integral equations which correspond to continuous signal systems. The Z transform may be used to express discrete "signals" in terms of sums of geometric sequences just as the Fourier series or integral expresses continuous "signals."

Many operations encountered in information processing may be expressed by Laplace, Mellin, and Hankel transforms. The Laplace transform has perhaps been developed more fully because of its relatively simplicity for the operations shown on Table F-2. The Mellin transform is appropriate for multiplication, differentiation, and convolution in the time domain while the Hankel transform is suitable for scale changes (compression or expansion) and for first order differentiation. It is important to note that a linear addition in one domain goes over as a linear addition in the conjugate domain for all three transforms, since all integral transforms are derivable from the superposition integral.

TABLE F-I
Effectiveness of Transforms for Solving Several Linear Differential, Integral, and Difference Equations Describing System Behavior

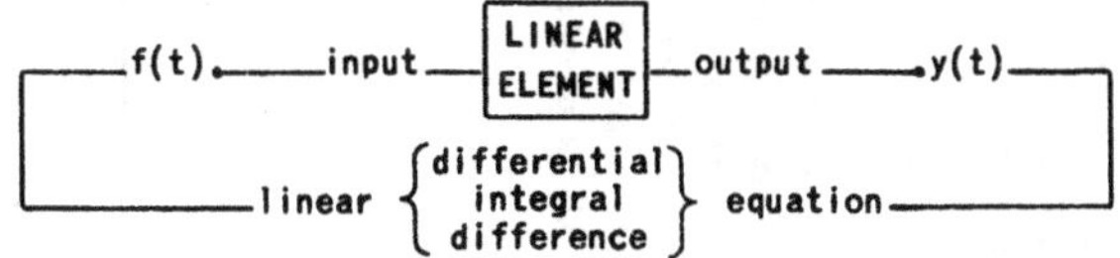

Equations Describing System Behavior		Solution Using Transforms	
Class	**Equations of Each Class**	**Type**	**Result of Transformation**
Differential	1. nth order differential equation with mth degree polynomial coefficients: $$\sum_{k=0}^{n}\sum_{r=0}^{m} a_{kr}t^r \frac{d^k y(t)}{dt^k} = f(t) \quad n > m$$	Laplace $$F(p) = \int_{0+}^{\infty} f(t)\, e^{-pt}\, dt$$	$$\sum_{k=0}^{n}\sum_{r=0}^{m} (-1)^r a_{kr} \frac{d^r}{dp^r}\left[p^k y(p) - \sum_{\ell=1}^{k} p^{k-\ell} \left.\frac{d^\ell y(t)}{dt^\ell}\right\|_{t=0}\right] = F(p)$$ An mth order differential equation having nth degree polynomial coefficients. For the special case of constant coefficients (r = 0), the above reduces to an algebraic equation.
	2. Euler-Cauchy differential equation: $$\sum_{n=0}^{N} a_n t^n \frac{d^n y(t)}{dt^n} = f(t)$$	Mellin $$F(s) = \int_{0}^{\infty} f(t)\, t^{s-1}\, dt$$	$$Y(s) = \frac{F(s)}{\sum_{n=0}^{N} a_n(-1)^n s(s+1)(s+2)\ldots(s+n+1)}$$ (Algebraic)
	3. The Bessel differential equation: $$\left[\frac{d^2}{dt^2} + \frac{1}{t}\frac{d}{dt} - \frac{n^2}{t^2} \pm a^2\right]^N y(t) = f(t)$$	Hankel $$F_n(\sigma) = \int_{0}^{\infty} tJ_n(\sigma t)\, f(t)\, dt$$	$$Y_n(\sigma) = \frac{F_n(\sigma)}{(\pm 1)^N (a^2 \mp \sigma^2)^{2N}}$$ (Algebraic)
	4. Differential equation where the time-varying parameter is a function of the first power of t. Example: The Laguerre equation $$t\frac{d^2y(t)}{dt^2} + (1-t)\frac{dy(t)}{dt} + ny = f(t)$$ $y(o) = 1$, $n = 0$ or an integer	Laplace $$F(p) = \int_{0+}^{\infty} f(t)\, e^{-pt} dt$$	$$Y'(p) + \frac{p-(n+1)}{p(p-1)} Y(p) = F(p)$$ The transform equation is a new differential equation of reduced order (1). For the special case of f(t) = 0, Y(p) is solved as: $$Y(p) = \frac{(p-1)^n}{p^{n+1}}$$ which is the Laplace transform of the Laguerre polynomial $$L_n(t) = \frac{1}{n!} e^t \frac{d^n}{dt^n}(t^n e^{-t})$$
Integral	1. The Volterra integral equation of the second kind with a difference kernel: $$\int_{0}^{t} k(t-\tau)\, y(\tau)\, d\tau + y(t) = f(t)$$	Laplace $$\begin{Bmatrix} F(p) \\ K(p) \end{Bmatrix} = \int_{0+}^{\infty} \begin{Bmatrix} f(t) \\ k(t) \end{Bmatrix} e^{-pt}\, dt$$	$$Y(p) = \frac{F(p)}{1 + K(p)}$$ (Algebraic)
	2. The Fredholm integral equation of the first type with a product kernel: $$\int_{0}^{t} k(t\tau)\, y(\tau)\, d\tau = f(t)$$	Mellin $$\begin{Bmatrix} F(s) \\ K(s) \end{Bmatrix} = \int_{0}^{\infty} \begin{Bmatrix} f(t) \\ k(t) \end{Bmatrix} t^{s-1}\, dt$$	$$V(s) = \frac{F(s)}{K(s)}$$ which is an algebraic equation. Once v(t) is obtained by inversion, y(t) can be obtained from the following relation: $$y(t) = \frac{v\left(\frac{1}{t}\right)}{t}$$
Difference	Linear difference equation: $$\alpha_0 y_n + \alpha_1 y_{n-1} + \alpha_2 y_{n-2} + \cdots = \beta_0 f_n + \beta_1 f_{n-1} + \beta_2 f_{n-2} + \cdots$$	"Z" $$F(z) = \sum_{n=0}^{\infty} f(nT)\, e^{-z}$$ T = sampling interval	$$Y(z) = \frac{F(z) \sum_{n=0}^{\infty} \beta_n z^{-n}}{\sum_{n=0}^{\infty} \alpha_n z^{-n}}$$ which is clearly algebraic. Calculating Y(z), the response y_n at time $t_n = nT$ is $$y_n = y(t_n) = Y(z)z^n$$

TABLE F-2
General Operations in System Analysis and Synthesis Together with Their Conjugate Operations in the Laplace, Mellin, and Hankel Domains

OPERATIONS ON FUNCTIONS	LAPLACE DOMAIN $L[f(t)] = F_L(s)$	MELLIN DOMAIN $M[f(t)] = F_M(s)$	HANKEL DOMAIN $H[f(t)] = F_{H_n}(s)$	CONCLUSIONS
Linear Addition $A\ f(t) + B\ g(t)$	$AF_L(s) + BG_L(s)$	$AF_M(s) + BG_M(s)$	$AF_{H_n}(s) + BG_{H_n}(s)$	Linear addition in one domain goes over as a linear addition in the conjugate domain for all three transforms. This is true for all integral transforms.
Delay $f(t - T)$	$e^{-sT} F_L(s)$			A delay in the time domain causes a linear increase in phase in the Laplace domain. There is no simple representation in the Mellin and Hankel domains.
Scale Change $f(Kt)$	$\frac{1}{K} F_L\left(\frac{s}{K}\right)$	$K^{-s} F_M(s)$	$\frac{1}{K} F_{H_n}\left(\frac{s}{K}\right)$	Time compression or expansion results in "frequency" expansion or compression, respectively, in Laplace and Hankel Domains. This is known as reciprocal spreading and does not occur in the Mellin domain where a scale change alters the shape of the transform.
Multiplication $f(t)g(t)$	$\frac{1}{2\pi j} \int_{c-j\infty}^{c+j\infty} F_L(s') G_L(s - s')\, ds'$	$\frac{1}{2\pi j} \int_{c-j\infty}^{c+j\infty} F_M(s') G_M(s - s')\, ds'$		Multiplication in the time domain may be represented as a convolution of the transforms of the functions in both the Laplace and Mellin domains, entailing complex integration. It is not easily represented in the Hankel domain.
Differentiation $\frac{df(t)}{dt}$	$sF_L(s) - f(0+)$	$-(s - 1) F_M(s - 1)$ (assuming $t^{s-1} f(t) = 0$ at $t = 0$ and $t = \infty$)	$-s\left[\frac{(n+1)}{2n} F_{H_{n-1}}(s) - \frac{(n-1)}{2n} F_{H_{n+1}}(s)\right]$ (assuming $tf(t) = 0$ at $t = 0$ and $t = \infty$)	First order differentiation is easily represented in all three domains. The one used would depend upon the type of function or system involved.
Convolution (a) $\int_0^t f(\tau)\, g(t - \tau)\, d\tau$ (b) $\int_0^t f(\tau)\, g\left(\frac{t}{\tau}\right) d\tau$	(a) $F_L(s)\ G_L(s)$	(b) $F_M(s)\ G_M(s)$		There is no simple convolution theorem for the Hankel domain as there is for the Laplace and Mellin domains. A convolution in the latter two domains results in a multiplication in the corresponding conjugate domain, depending upon the type of convolution involved.

REFERENCES

F-1. E. A. Guillemin, "The Mathematics of Circuit Analysis," John Wiley and Sons, Inc., 1956, Chapter VII

F-2. S. A. Schelkunoff, "Applied Mathematics for Engineers and Scientists," D. Van Nostrand Company, Inc., 1948, Chapter II

F-3. R. M. Lerner, "The Representation of Signals," I.R.E. Trans., Vol. CT-6, May 1959, pp. 197-216

F-4. G. Szego, "Orthogonal Polynomials," Am. Math. Soc., 1959

F-5. R. Courant and D. Hilbert, "Methods of Mathematical Physics," Volume I, Interscience Publishers, Inc., 1953, Chapters I and II

F-6. J. A. Aseltine, "Transform Method in Linear System Analysis," McGraw-Hill Book Company, Inc., 1948

F-7. V. Volterra, "Theory of Functionals and of Integral and Integrodifferential Equations," Blackie, London, 1930

F-8. R. V. Churchill, "Introduction to Complex Variables and Applications," McGraw-Hill Book Company, Inc., 1948

F-9. W. T. Thomson, "Laplace Transformation Theory and Engineering Applications," Prentice-Hall, New York, 1950

F-10. I. N. Sneddon, "Fourier Transforms," McGraw-Hill Book Company, Inc., 1948

F-11. D. V. Widder, "The Laplace Transform," Princeton University Press, 1941

F-12. E. C. Titchmarch, "Introduction to the Theory of Fourier Integrals," Oxford University Press, London, 1948, Chapter 8

F-13. G. A. Cambell and R. M. Foster, "Fourier Integrals for Practical Applications," Van Nostrand Company, Inc., 1940

F-14. A. Erde'lyi, "Tables of Integral Transforms," McGraw-Hill Book Company, Inc., 1954, Vol. 2, Chapter 10

F-15. C. J. Tranter, "Integral Transforms in Mathematical Physics," John Wiley and Sons, Inc., 1951

F-16. F. R. Gerardi, "Application of Mellin and Hankel Transforms to Networks with Time-Varying Parameters," I.R.E. Trans., Vol. CT-6, No. 2, June 1959, pp. 197-208

F-17. J. T. Tou, "Digital and Sampled-Data Control Systems," McGraw-Hill Book Company, Inc., 1959, Chapter 5

F-18. W. V. Lovitt, "Linear Integral Equations," Dover Publications Inc., New York, 1950

G. LINEAR ANALYSIS OF CIRCUIT ELEMENTS

1. INTRODUCTION

The behavior of linear elements can be described by integro-differential equations involving the input and output. Such equations relate the output to the input and formulate the basic relationship between "cause and effect," the exact nature of which depends upon the specific element. When the element is linear, the integro-differential equation will be of the form

$$\begin{aligned} a_n(t)\frac{d^n r(t)}{dt^n} + a_{n-1}(t)\frac{d^{n-1} r(t)}{dt^{n-1}} + \cdots + \underbrace{\int\cdots\int}_{m\text{-fold}} a_m(t)\, r(t)dt_1 \cdots dt_m + \cdots + a_o(t)r(t) = \\ b_v(t)\frac{d^v c(t)}{dt^v} + b_{v-1}(t)\frac{d^{v-1} c(t)}{dt^{v-1}} + \cdots + \underbrace{\int\cdots\int}_{p\text{-fold}} b_p(t)\, c(t)dt_1 \cdots dt_p + \cdots + b_o(t)c(t) \end{aligned} \qquad \text{(G-1)}$$

where $c(t)$ is the driving function and $r(t)$ is the output. The linearity is due to the fact that the dependent variable and its derivatives and integrals, are of the first power and are combined in a linear equation.

The basic property of linear elements is that of linear superposition. It is important to recognize that linearity does not require that coefficients of the integro-differential equation be constants but that their values be independent of either the input or output. If the coefficients are constant, the element is linear time-invariant and has the property that when an element's response to a given function is known, the response to the derivative, or integral of the input function may be found by differentiating, or integrating the original response. If the coefficients are independent functions of time, the element is linear time-variant and the analysis is more complex. Sections G-1 to G-11 will be concerned with the characterization of linear time-invariant elements, and in section G-12, the analysis is extended to include linear time-varying elements. The reason for discussing linear elements is that most elements can be considered to be linear at least over some operating range. The nonlinearities may then be in terms of second order perturbation effects.

2. LINEAR SUPERPOSITION

The concept of linear superposition plays an extremely important role in element analysis and synthesis. A basic aspect of the superposition principle is the classical method of obtaining the complete solution of a linear differential equation. This is performed by taking the sum of the free solution (transient response) and of the forced solution (steady-state response) where the initial conditions are used to evaluate the constants of integration. This technique may be extended by resolving any general source function into component functions for which the solutions can be found more readily.

A useful form of the superposition theorem utilizes the response of an element to a step source function. If a general source function $c(t)$ is resolved into step source components as is illustrated in Figure (G-1), then the response $r(t)$ of an element to this excitation may be given in terms of its normalized response $A(t)$ to a step source, defined as

$$A(t) = \frac{\text{response to step source}}{\text{magnitude of step source}} \qquad \text{(G-2)}$$

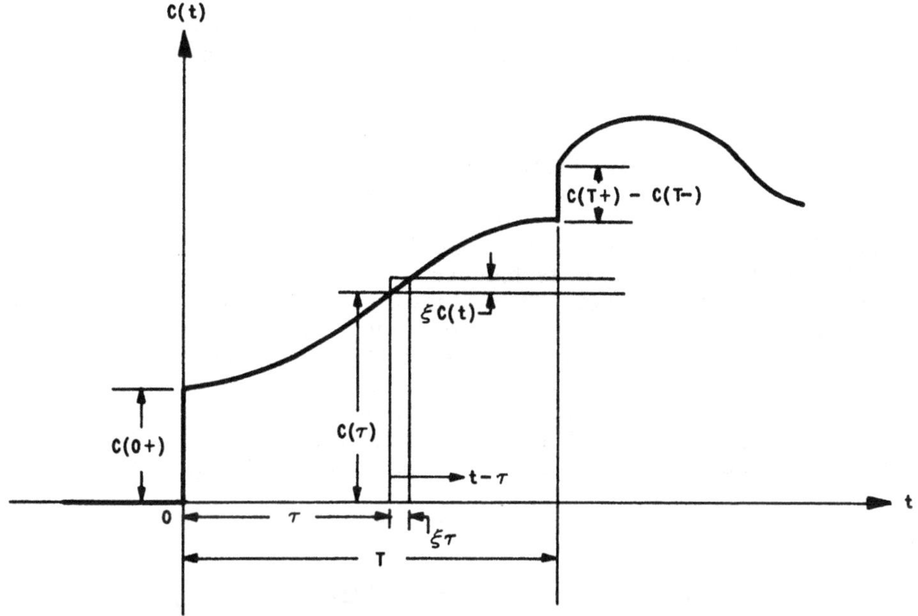

Figure G-1 - Resolution of general source function into step source components

If $c(t)$ is a continuous function of time t from $0 < t < \infty$ and its step source components are increased without bound, then the summation of the responses to these components may be approximated by an integral:

$$r(t) = A(t)\, c(o+) + \int_{\tau=0}^{\tau=t} A(t-\tau) \left. \frac{dc(t)}{dt} \right|_{t=\tau} d\tau \, . \tag{G-3}$$

This is often referred to as the Duhamel integral or superposition integral. If the excitation $c(t)$ has finite discontinuities as in Figure (G-1) at $t = T$, then the integral in (G-3) must be separated into those parts where $c(t)$ is continuous over the limits of integration and a response term to distinguish the finite discontinuity. Therefore, if $t > T$, then

$$r(t) = A(t)\, c(o+) + \int_{\tau=0}^{\tau=t} A(t-\tau) \left. \frac{dc(t)}{dt} \right|_{t=\tau} d\tau$$

$$+ A(t-T) \left[c(T+) - c(T-) \right] + \int_{\tau=T}^{\tau=t>T} A(t-T) \left. \frac{dc(t)}{dt} \right|_{t=\tau} d\tau \, . \tag{G-4}$$

Equation (G-3) may also be expressed as:

$$r(t) = \frac{d}{dt} \int_{\tau=0}^{\tau=t} A(t-\tau)\, c(\tau)\, d\tau \, . \tag{G-5}$$

This form may prove more useful depending on the complexity of the derivative of $c(t)$ and of the integral.

Thus, if the response of an element to a step source is known, then its response to an arbitrary source function can be deduced from this information. In general, any linear cause and effect relationship involving a single independent variable can be expressed in the form of a modified superposition theorem such as (G-3) or (G-5). An indication of the importance of the superposition integral is that all integral transforms may be derived from it.

3. TRANSFER FUNCTION

Previously it had been indicated that the representation of functions could be made by a large number of methods. A variety of techniques are also available for describing physical elements having an input and an output. Some of these include the use of circuit diagrams or mathematical equations. One method for describing the characteristics of a linear, stationary network is to specify its transfer function. The transfer function is defined to be the ratio of the Fourier (or Laplace) transforms of the output and input. The important characteristics of a transfer function establish the dependence of certain of its properties as a function of frequency, such as gain or phase angle. Thus, the complex transfer function $H(j\omega)$ of an element may be described by steady-state transmission properties, for example,

$$H(j\omega) = e^{A(j\omega)} e^{jB(j\omega)} \tag{G-6}$$

$$= |H(j\omega)| \underline{/B(j\omega)} \tag{G-7}$$

where $A(j\omega) = \log |H(j\omega)|$ is the gain, and $B(j\omega) = \text{angle } H(j\omega)$ is the phase.

The transfer function may be determined by measurement or by analysis. The most commonly used measuring technique is to apply a sinusoidal input to the element and record the output amplitude and phase. This test is repeated at a number of frequencies to determine the gain and phase curves. Obtaining the transfer function by analysis is accomplished by an algebraic process that is equivalent to solving the element's differential equation for the steady-state sinusoidal input case. The procedure for such an analysis is the following:

(1) Determine the differential equation relating output to input.

(2) Substitute the algebraic term $(j\omega)$ for the operation d/dt, $(j\omega)^2$ for d^2/dt^2, $1/j\omega$ for $\int dt$, $1/(j\omega)^2$ for $\int\int(dt)^2$.

(3) Solve the resultant equation for the ratio of output to input as a function of radian frequency ω.

(4) Convert the ratio from complex form to magnitude-phase form and plot the results.

If the input is a periodic time function $f_i(t)$ whose Fourier series converges, then

$$f_i(t) = \sum_{n=-\infty}^{\infty} D(jn\omega_o)\, e^{jn\omega_o t} \qquad T = \frac{2\pi}{\omega_o} \tag{G-8}$$

where T is the fundamental period, and the complex coefficients $D(jn\omega_o)$ are given by

$$D(jn\omega_o) = \frac{1}{T} \int_{-T/2}^{T/2} f_i(t)\, e^{-jn\omega_o t}\, dt\,. \tag{G-9}$$

From the definition of the transfer function $H(j\omega)$, the system output in response to the nth input component is

$$f_{o_n}(t) = D(jn\omega_o)\, H(jn\omega_o)\, \exp\,(jn\omega_o t)\,. \tag{G-10}$$

That is, the output is the input component, its amplitude being multiplied by the absolute magnitude of the transfer function at the frequency $n\omega_o$ and the phase advanced by the angle of

$H(jn\omega_o)$, the phase of the element at this frequency. Since the system is linear, its total response to $f_i(t)$ is the sum of the component outputs $f_{o_n}(t)$ and may be represented by

$$f_o(t) = \sum_{n=-\infty}^{\infty} D(jn\omega_o)\, H(jn\omega_o)\, \exp(jn\omega_o t)\,. \tag{G-11}$$

This is the steady-state output of a linear, constant-coefficient element when its input is a periodic function of time as given by (G-8).

From Fourier analysis, a knowledge of $H(j\omega)$ for all frequencies determines the transient as well as the steady-state properties of linear filters. If $f_i(t)$ is a transient having a Fourier transform $F_i(j\omega)$, then the Fourier transform of the output of a linear, stationary device in response to the transient input is equal to the Fourier transform of the input times the transfer function, i.e.,

$$F_o(j\omega) = F_i(j\omega)\, H(j\omega)\,. \tag{G-12}$$

4. IMPULSE RESPONSE

Another characterization of an element is in terms of time functions. A very useful and important time function is the unit impulse $\delta(t)$ defined by the relations

$$\delta(t-T) = 0 \qquad t \neq T \tag{G-13}$$

$$\int_{T-\epsilon}^{T+\epsilon} \delta(t-T)\, dt = 1 \qquad \epsilon > 0\,. \tag{G-14}$$

Equation (G-13) indicates that $\delta(t-T)$ is zero everywhere except at $t=T$ while (G-14) requires that the impulse function have unit area. The Fourier spectrum of this input is

$$F_i(j\omega) = \int_{-\infty}^{\infty} \delta(t)\, e^{-j\omega t}\, dt = 1. \tag{G-15}$$

Thus, the spectrum of the unit impulse has unit amplitude and zero phase for all frequencies. It is interesting to note that the energy contained in a unit impulse in any frequency band $\omega_1 < \omega < \omega_2$ is proportional to the bandwidth, i.e.,

$$\int_{\omega_1}^{\omega_2} |F_i(\omega)|^2\, d\omega = \int_{\omega_1}^{\omega_2} d\omega = \omega_2 - \omega_1\,. \tag{G-16}$$

The energy of the frequency components in a unit impulse is therefore concentrated at the extremely high frequencies.

By substituting (G-15) in (G-12)

$$F_o(j\omega) = 1 \cdot H(j\omega) = H(j\omega)\,. \tag{G-17}$$

Thus, the Fourier spectrum of the output of an element when a unit impulse is applied is the transfer function of the element. The output $f_o(t)$ is expressed as

$$f_o(t) = \frac{1}{2\pi}\int_{-\infty}^{\infty} F_o(j\omega)\, e^{j\omega t}\, d\omega = \frac{1}{2\pi}\int_{-\infty}^{\infty} H(j\omega)\, e^{j\omega t}\, dt\,. \tag{G-18}$$

The unit impulse response (commonly designated $h(t)$) of a fixed-parameter linear element is therefore given by the Fourier transform of the transfer function $H(j\omega)$ of that element. Since

$$H(j\omega) = \int_{-\infty}^{\infty} h(t)\, e^{-j\omega t}\, dt, \qquad \text{(G-19)}$$

then similar to the case for a step source response, the response of an element to an arbitrary source function can be obtained if its response to an impulse is known.

5. CONVOLUTION THEOREM

In the discussion of the superposition integral it was indicated that the response $r(t)$ of a fixed-parameter linear element to an arbitrary input $c(t)$ could be expressed in terms of the element's step response $A(t)$; one such relationship was given by (G-5), as follows:

$$r(t) = \int_{0}^{t} \frac{d}{dt} A(t-\tau)\, c(\tau)\, d\tau. \qquad \text{(G-5)}$$

The derivative operator has been brought inside the integrand for convenience. For time-invariant elements the response to the derivative of an input signal is the derivative of the response to the original signal. Since the derivative of a step function is an impulse, the derivative with respect to time of the displaced step response $A(t-\tau)$ in (G-5) may be considered as the displaced impulse response $h(t-\tau)$. The excitation $c(\tau)$ is unaffected by the operation since it is a function of τ only. (G-5) may then be written as

$$r(t) = \int_{0}^{t} h(t-\tau)\, c(\tau)\, d\tau. \qquad \text{(G-20)}$$

This relationship is referred to as the convolution integral and is equivalent to representing the input function by a series of weighted impulses and then summing the responses to each. The convolution integral is a fundamental tool in linear analysis and contains the condition for the physical realizability of a transfer function, namely, that the input and output be related by (G-20).

6. TRANSFER FUNCTION METHOD

In referring to elements as being linear, it should be recognized that there are several properties of a signal to which the element can respond linearly, for example, it may be linear on the basis of amplitude, or on the basis of energy. In optics, the terms "coherent" and "incoherent" elements are used to apply to elements which respond linearly to the amplitude and to the energy of the input, respectively. These terms may also be applied to circuit elements.

When an element is coherent, having a transfer function $H(j\omega)$, then by applying Fourier analysis the frequency spectrum of the output $F_o(j\omega)$ can be given as

$$F_o(j\omega) = H(j\omega)\, F_i(j\omega) \qquad \text{(G-12)}$$

where $F_i(j\omega)$ is the Fourier transform of the input. It is seen that with a coherent element both the amplitude and phase of the Fourier components of the input may be controlled and that the attribute of the Fourier method is that a convolution integral is replaced by a multiplication process.

For a stable linear incoherent system the following relationship between input (stochastic) and output holds:

$$|F_o(j\omega)|^2 = |H(j\omega)|^2 \, |F_i(j\omega)|^2 \tag{G-21}$$

This states that the power spectrum of the output of an incoherent element is equal to the power spectrum of the input times the square of the magnitude of the transfer function. An incoherent element is insensitive to phase information and behaves as an envelope detector. A relationship such as (G-21) will exist whenever the input is random noise.

7. USE OF RANDOM NOISE IN DETERMINING TRANSFER FUNCTIONS

If $x(t)$, a typical member of an ergodic ensemble, is the input to a fixed-parameter stable linear system, then the output $y(t)$ will also be a typical member of an ergodic ensemble. The cross-correlation function $\psi_{xy}(\tau)$ between input and output is

$$\psi_{xy}(\tau) = \lim_{T \to \infty} \frac{1}{2T} \int_{-T}^{T} x(t)\, y(t+\tau)\, dt \,. \tag{G-22}$$

The output may also be expressed by the convolution theorem

$$y(t) = \int_0^{\infty} h(\beta)\, x(t-\beta)\, d\beta \,. \tag{G-23}$$

Substituting (G-23) in (G-22) and changing the order of integration:

$$\psi_{xy}(\tau) = \int_0^{\infty} h(\beta)\, \psi_{xx}(\tau-\beta)\, d\beta \tag{G-24}$$

where ψ_{xx} is the autocorrelation function of the input. Equation (G-24) is a fundamental equation for any linear transmission system whose input is a typical function of an ergodic ensemble. Taking the Fourier transform of (G-24)

$$W_{xy}(f) = H(j\omega)\, W_{xx}(f) \tag{G-25}$$

$$H(j\omega) = \frac{W_{xy}(f)}{W_{xx}(f)} \,. \tag{G-26}$$

Thus, the transfer function is equal to the ratio of the cross-power spectrum between input and output to the power spectrum of the input. Since $W_{xx}(f)$ is real, the phase of $H(j\omega)$ is the same as that of $W_{xy}(f)$ for all frequencies.

For the special case of white noise, the power spectrum is a constant

$$W_{xx}(f) = N_o \,. \tag{G-27}$$

The transfer function $H(j\omega)$ and impulse response $h(\tau)$ then simplify to

$$H(j\omega) = \frac{W_{xy}(f)}{N_o} \tag{G-28}$$

$$h(\tau) = \frac{\psi_{xy}(\tau)}{N_o} \,. \tag{G-29}$$

8. APPLICATIONS OF FOURIER ANALYSIS

Fourier analysis is most useful in expressing the transmission properties of linear time-invariant elements. This is attributed to the fact that a Fourier frequency decomposition is insensitive to translations in time (only the phase of the Fourier spectrum is modified). It has been previously shown that Fourier analysis can be extended to nonrepetitive as well as repetitive signals and consequently has many applications in element design. Some of these are:

(a) To predict the element's response,

(b) To determine the element's dynamic specifications,

(c) To evaluate or interpret test results.

In determining the output of an element for a given input, it is desirable to know its response to a sinusoid. If the transfer function is $H(j\omega)$ and a sine wave of angular frequency ω_o and amplitude (E) is applied to the input, the output will also be a sine wave of frequency ω_o, having an amplitude $|H(j\omega_o)| \cdot E$ with its phase advanced by the angle of $H(j\omega_o)$. By resolving an arbitrary input into its harmonic components, each a sinusoidal wave of different frequency, the corresponding outputs may be determined. For a linear element, the resultant response is the complex addition of these outputs. This method is illustrated in Figure (G-2) for a square-wave input and a given element.

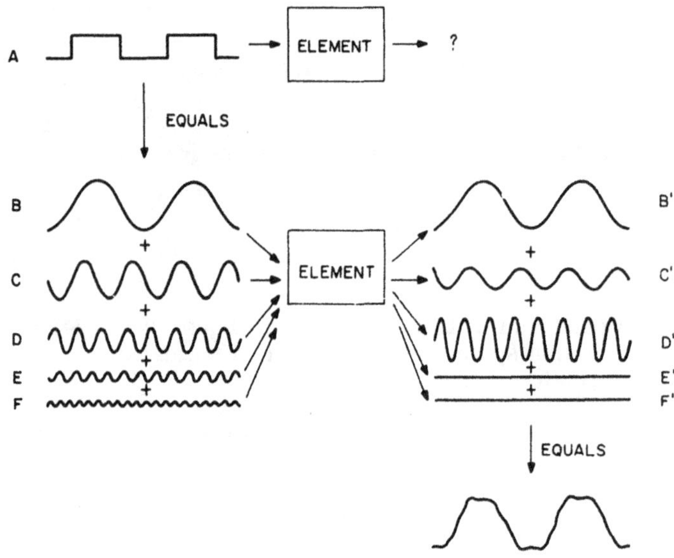

Figure G-2 - Fourier analysis to predict element response

To determine the gain and phase characteristics of an element which are necessary to produce a desired output for a given input, the input and output waveforms are resolved into their respective Fourier components and compared at corresponding frequencies. This provides amplitude and phase information, two sets of requirements which the element must satisfy. When the decomposition yields harmonic components at different frequencies, the element is nonlinear. Figure (G-3) illustrates the procedure where reference time t_o is needed to determine phase requirements.

The preceding method may also be used to determine the gain and phase characteristics of an unknown element from known test results. It should be recognized, however, that in so doing it is assumed that the element is linear.

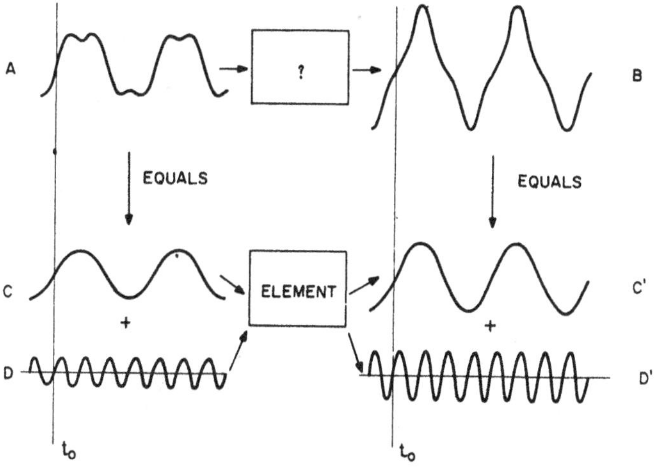

Figure G-3 - Fourier analysis to determine element's dynamic specifications

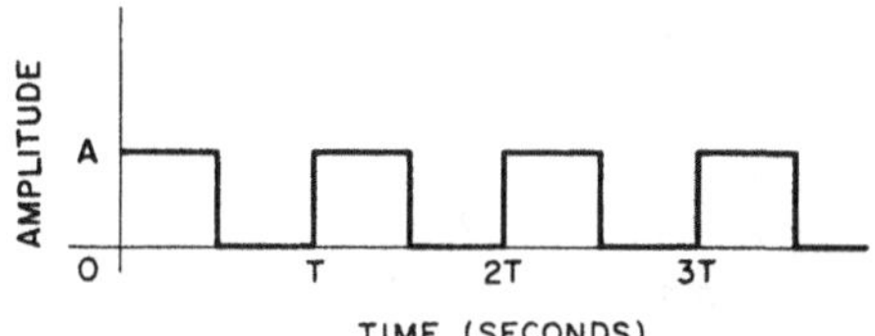

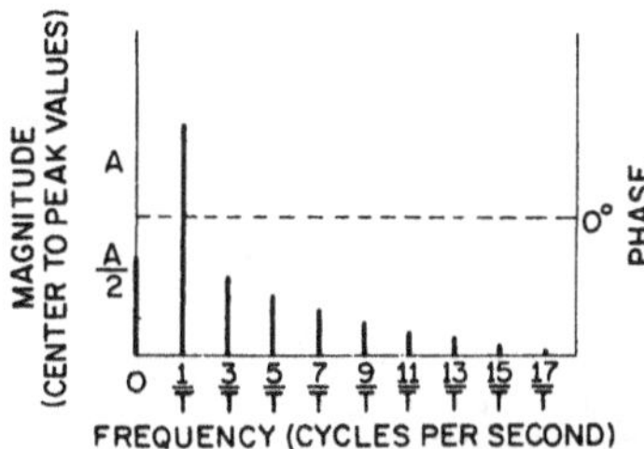

Figure G-4 - Harmonic components of a displaced square wave

In making a Fourier analysis, it is necessary to interrelate the time and frequency domains, since the effects of errors made in one domain must be known in the transform domain. Representations in time and frequency for a square wave are shown in Figure (G-4). The discussion of linear superposition in section (G-2) showed that an element's response to a step input may be used to determine the output waveshape for an arbitrary input. Practically, this cannot be done since a step function has many significant harmonic components. Consequently, the output cannot be easily calculated unless an approximation is made. Comparing Figures (G-5.a) and (G-5.b), it is seen that the step response is the same as the first half-cycle of the output response to a displaced square-wave input. Therefore, the readily calculated response of Figure (G-5.a) gives the desired response of Figure (G-5.b) during the first t_o seconds. This technique is used to find the step response of any system whose transfer function is given. The validity of the results and facility of calculation will depend on the period chosen for the square wave. If too short, as in Figure (G-5.c), the output will never reach steady-state and if too long, as in Figure (G-5.d), there will be an unnecessarily large number of harmonic components to calculate. A suitable frequency for the square-wave cycle is about one eighth the cut-off frequency of the transfer function. The fundamental and the odd harmonics up to the twenty-fifth should be calculated for a reasonable approximation, making fourteen harmonic terms in all.

To determine the steady-state response of an element from the transient response, it is assumed the element behaves linearly throughout the transient response test. The test data should be known with sufficient accuracy since small changes in the transient response may be equivalent to larger changes in the frequency response. The procedure consists of first replacing the step input with a train of pulses of equal amplitude. Then the response to the pulses will be the same as for the step for the duration of the pulse. At this time, the input and output waveforms are assumed to return to zero in the same manner as they originated

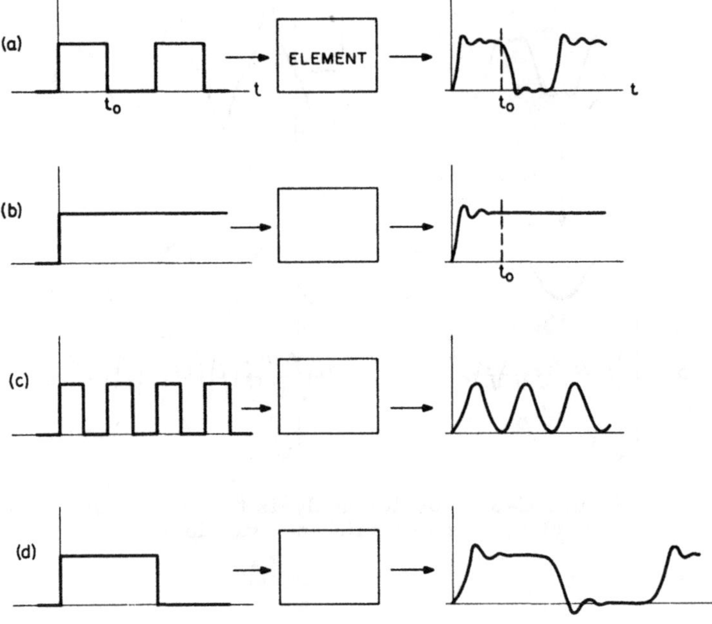

Figure G-5 - Determining transient response from frequency response

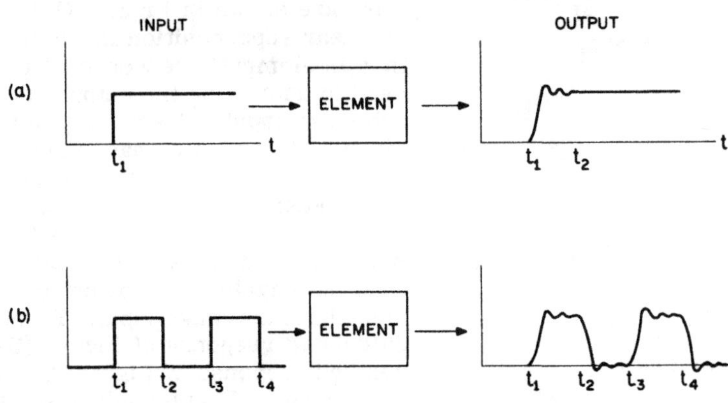

Figure G-6 - Determining frequency response from transient response

and are repeated in alternate inversions in equal time periods. This is illustrated in Figure (G-6). The transfer function is determined using the method described earlier and illustrated in Figure (G-3).

9. GENERAL CONDITIONS FOR PHYSICAL REALIZABILITY AND STABILITY

Criteria for physical realizability can be given in terms of either the impulse response $h(t)$ or the transfer function $H(j\omega)$. Specifying one implicitly defines the other since they are related through the Fourier transform.

Use of the impulse response involves the following requirements:

(1) $h(t)$ must be zero for $t < 0$,

(2) $h(t)$ must approach zero (with reasonable rapidity) as t approaches $+\infty$.

The first condition states that a network cannot respond to an impulse before the impulse arrives, while the second implies that the effect of an impulse will eventually die out. The latter thus ensures the correlation introduced by a realizable element to be of finite range so that if the input to such an element was ergodic, the output would also be an ergodic function. These conditions are sufficient as well as necessary for physical realizability. Sufficiency is used in the sense that any impulsive response $h(t)$ satisfying both conditions can be approximated as closely as desired with a passive linear network, used with an ideal amplifier.

In terms of the frequency response, the principal conditions for physical realizability are referred to $H(j\omega)$, considered as a function of the complex variable ω. $H(j\omega)$ must:

(1) be an analytic function in the half-plane defined by $\mathrm{Im}(\omega) < 0$,

(2) behave on the real frequency axis such that

$$\int_0^\infty \frac{\log |H(j\omega)|}{1 + \omega^2} \, d\omega$$

is a finite number.

The first condition establishes stability, that is, the element must not be capable of an oscillation that builds up in time. The second condition specifies the requirement for the amplitude function.

If $H(\omega)$ is a transfer function satisfying both conditions, then for a given gain function $A(\omega) = \log |H(\omega)|$ there is a minimum possible phase characteristic. For a network of the minimum phase type, $H(\omega)$ has neither zeros nor poles in the half plane defined by $\mathrm{Im}(\omega) < 0$. The phase $B(\omega_o)$ at the frequency $(\omega_o/2\pi)$ is given by

$$B(\omega_o) = \frac{2\omega_o}{\pi} \int_0^\infty \frac{A(\omega) - A(\omega_o)}{\omega^2 - \omega_o^2} \, d\omega . \qquad \text{(G-30)}$$

If the derivative of the gain function is easier to work with than the gain function itself, then (G-30) may be expressed as

$$B(\omega_o) = \frac{1}{\pi} \int_0^\infty \frac{dA}{d\omega} \log \left| \frac{\omega + \omega_o}{\omega - \omega_o} \right| d\omega . \qquad \text{(G-31)}$$

A minimum phase element has the important property that its inverse, with the transfer function $H^{-1}(\omega)$, is also physically realizable. A signal passed through a minimum phase element $H(\omega)$ may be recovered by passing its output through an inverse element $H^{-1}(\omega)$ without incurring a time delay. If a signal is transmitted through a nonminimum phase element, the best that can be done is to provide an element having the properties of the theoretical inverse except for a phase lag. Thus, there is no physically realizable exact inverse for a nonminimum phase element and a signal passed through it can be recovered only after a delay. Both types of elements, including the inverse of the minimum phase network, are illustrated in Figure (G-7).

All physically realizable impulse responses and transfer functions may be approximated with desired accuracy with passive lumped linear networks and ideal active elements. A passive linear network, when defined in terms of time and energy, is restricted by the following conditions:

(1) it is linear,

(a) Filter of minimum phase shift type

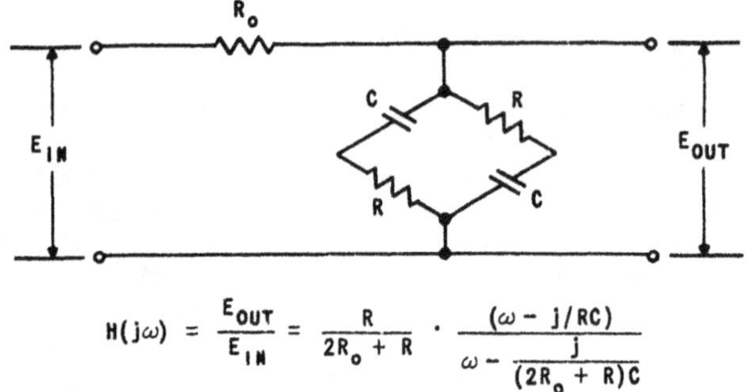

$$H(j\omega) = \frac{E_{OUT}}{E_{IN}} = \frac{R}{2R_0 + R} \cdot \frac{(\omega - j/RC)}{\omega - \dfrac{j}{(2R_0 + R)C}}$$

(b) Filter of non-minimum phase shift type

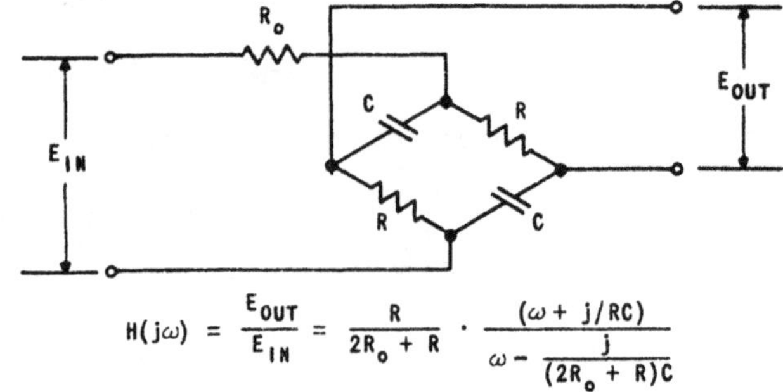

$$H(j\omega) = \frac{E_{OUT}}{E_{IN}} = \frac{R}{2R_0 + R} \cdot \frac{(\omega + j/RC)}{\omega - \dfrac{j}{(2R_0 + R)C}}$$

(c) Filter of minimum phase shift type. If this is followed by an ideal amplifier of gain $(2R_0 + R)\ (r + r_1)/r_1R$, the combination is the inverse cf (a), provided $C_1 = 2R_0CC'/(RC - rC')$ and $r_1 = (2R_0 + R)(RC - rC')/2R_0C'$.

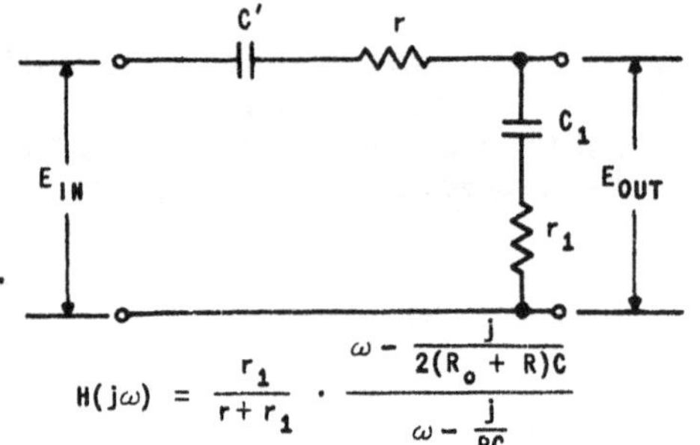

$$H(j\omega) = \frac{r_1}{r + r_1} \cdot \frac{\omega - \dfrac{j}{2(R_0 + R)C}}{\omega - \dfrac{j}{RC}}$$

Figure G-7 - Illustration of filters of (a) minimum phase shift type, (b) non-minimum phase shift type, and (c) the inverse of the minimum phase shift type

(2) the total energy at the network is positive,

(3) there is no response between any pair of terminals before an excitation is fed to the network.

For two-terminal networks, a necessary and sufficient condition that it be linear passive is that its impedance function be positive and real. From a consideration of the physical realizability of a lumped linear stable element the transfer function has several important properties:

(1) the transfer functions are expressed as ratios of polynomials of $j\omega$ (or s) with real coefficients,

(2) the numerator polynomial cannot be of higher degree than the denominator,

(3) the denominator polynomial can have roots in the left half-plane only.

10. SHORT-TIME AUTOCORRELATION FUNCTION AND POWER SPECTRUM

Ordinarily, the representations of autocorrelation functions show the averaging process over an infinite length of time. This implies, for example, that in the frequency domain, the passband of the filter with which the power spectrum is measured should be infinitely narrow. Both requirements are, of course, not realizable. The autocorrelation function and its corresponding power spectrum, when determined experimentally, are both inherently time dependent and consequently are approximations. For some applications it may be necessary to represent the correlation function or power spectrum in terms which correspond more closely to the conditions under which they have been determined experimentally.

Wiener's theorem relating the power spectrum to the autocorrelation function may be derived for finite time constants. The first step involves establishing physically realizable measuring procedures in terms of mathematical operations. The measurement of the short-time autocorrelation function may be defined by the following operations:

(a) The input function $f(t)$ is delayed by a time τ, yielding the function $f(t-\tau)$,

(b) Multiplication of $f(t)$ by $f(t-\tau)$, yielding the product function

$$\psi_\tau(t) = f(t)\, f(t+\tau) \tag{G-32}$$

(c) The function $\psi_\tau(t)$ is averaged by means of a lowpass filter having a transfer function

$$H(j\omega) = \frac{1}{1 + j(\omega/2a)} \tag{G-33}$$

where ω is the angular frequency. The output of the filter yields a point of the short-time autocorrelation function and may be expressed, using the convolution integral, as a weighted average of the whole past of the function $\psi_\tau(t)$, i.e.,

$$\psi_t(\tau) = \int_{-\infty}^{t} \psi_\tau(x)\, h(t-x)\, dx \tag{G-34.a}$$

$$= 2a \int_{-\infty}^{t} \psi_\tau(x)\, e^{-2a(t-x)}\, dx\,. \tag{G-34.b}$$

The weighting function is the impulse response of the low-pass filter, and is illustrated in Figure (G-8(a)).

The measurement of the short-time power spectrum is defined by the following operations:

(a) The input function $f(t)$ is passed through a bandpass filter having the transfer function,

$$H'(j\omega) = \frac{(2a)^{1/2}(\beta + j\omega)}{(\beta + j\omega)^2 + \omega_o^2} \tag{G-35}$$

where ω_o is the natural frequency of oscillation of the filter, and β is the damping constant. Let the output of the filter be $g'_\omega(t)$.

(b) The input function $f(t)$ is passed through another bandpass filter with the transfer function

$$H''(j\omega) = \frac{(2a)^{1/2}\,\omega_o}{(\beta + j\omega)^2 + \omega_o^2}\,. \tag{G-36}$$

(a)

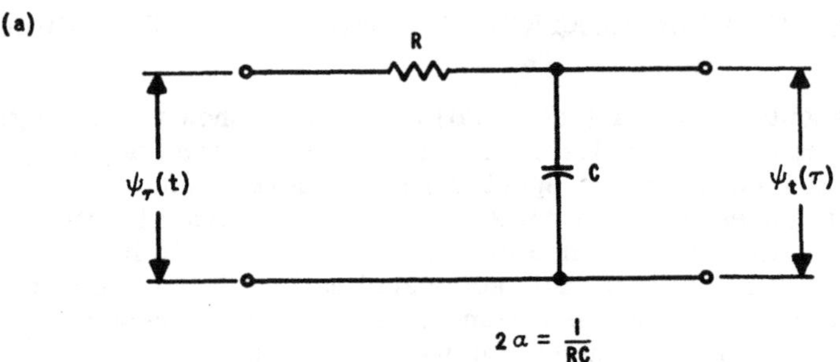

(b)

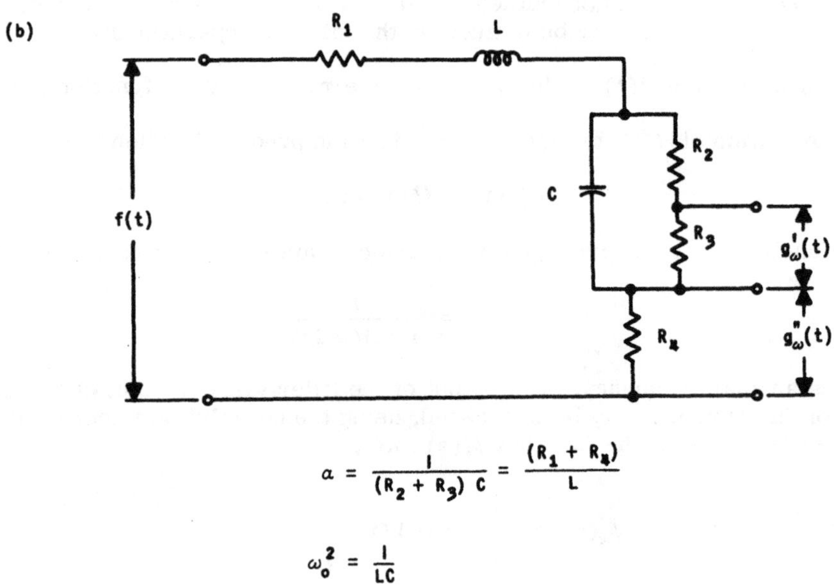

Figure G-8 - Networks yielding the transfer functions necessary for obtaining the short-time (a) autocorrelation function, and (b) power spectrum

The output of this second filter is designated as $g''_\omega(t)$.

(c) The outputs of the two filters are squared and added to yield a point of the short-time power spectrum

$$W_t(\omega) = \left[g'_\omega(t)\right]^2 + \left[g''_\omega(t)\right]^2 . \tag{G-37}$$

The transfer functions given in (G-35) and (G-36) may be realized by the network of Figure (G-8(b)).

The short-time autocorrelation function $\psi_t(\tau)$ is related to the short-time power spectrum $W_t(\omega)$ by

$$\psi_t(\tau) = \frac{e^{a|\tau|}}{2\pi} \int_{-\infty}^{\infty} W_t(\omega) \cos \omega\tau \, d\omega \tag{G-38}$$

and

$$W_t(\omega) = \int_{-\infty}^{\infty} \psi_t(\tau)\, e^{-a|\tau|} \cos \omega\tau \, d\tau \tag{G-39}$$

$$= \frac{1}{2\pi} \int_{-\infty}^{\infty} H_t(z) \frac{2a}{a^2 + (\omega - z)^2}\, dz \tag{G-40}$$

where

$$H_t(\omega) = \int_{-\infty}^{\infty} \psi_t(\tau) \cos \omega\tau \, d\tau \tag{G-41}$$

$$\frac{2a}{a^2 + \omega^2} = \int_{-\infty}^{\infty} e^{-a|\tau|} \cos \omega\tau \, d\tau \, . \tag{G-42}$$

Equation (G-41) indicates that $H_t(\omega)$ is the Fourier transform of the short-time autocorrelation function $\psi_t(\tau)$. Since $W_t(\omega)$ is a weighted average of $H_t(\omega)$, it seems to indicate that if the same value of a is used in both types of measurement, the short-time autocorrelation function will provide more accurate information about the power spectrum than the direct determination of the power spectrum by means of filters. The validity of the pair of reciprocal relations expressed by (G-38) and (G-39) is strictly limited to the results of the measuring procedures specified above. While other physical networks may be substituted for those shown in Figure G-8, only the latter yields an averaging process independent of τ and results that are reciprocally related at any time t.

The concept of a short-time power spectrum has been used in speech analysis where it is referred to as a sound spectrogram. Similarly, short-time autocorrelation functions of speech have been studied. The fact that these two tools of analysis are mathematically related increases their usefulness as representations.

11. LINEAR SYSTEM APPROXIMATION

INTRODUCTION

The approximation problem of linear system synthesis is the determination of a realizable system function which closely approximates a prescribed system function. If the latter is the impulse response $h(t)$, the approximate impulse response $h^*(t)$ is then defined by a sum of predetermined approximating functions $\phi_n(t)$ as

$$h^*(t) = \sum_{n=1}^{N} a_n \phi_n(t) \, . \tag{G-43}$$

As was indicated in the discussions on orthogonality, one way of specifying $h^*(t)$ is to select a_n such that the integral of the square of the magnitude of the difference between $h(t)$ and $h^*(t)$ is a minimum. This requires that the $\phi_n(t)$ be orthogonal, and if normalized, that they satisfy the relation

$$\int_{-\infty}^{\infty} \phi_n(t)\, \phi_m(t)\, dt = 1 \qquad n = m \tag{G-44}$$

$$= 0 \qquad n \neq m \, . \tag{G-45}$$

The minimum integral-square error $\mathcal{E}_{min}$ is then,

$$\mathcal{E}_{min} = \int_{-\infty}^{\infty} h^2(t)\,dt - \sum_{n=1}^{N} a_n^2 . \qquad \text{(G-46)}$$

Equation (G-46) requires that

$$\int_{-\infty}^{\infty} h^2(t)\,dt < \infty .$$

Realizability of $h^*(t)$ requires that $h^*(t)$, and consequently the $\phi_n(t)$, should be zero for negative time.

INTEGRAL-SQUARE ERROR

The integral-square error criterion is the basis of an orthonormal function approximation, resulting in an approximation error which generally oscillates about zero with relatively constant peak amplitude. A disadvantage of the square weighting is that there is no time interval of appreciable length where the error is very small. However, this disadvantage may be relieved by properly weighting the approximation (G-43).

Approximations in the time domain involve approximations, or errors, in the frequency domain. An acceptable approximation of the impulse response will ensure that the approximation of the gain, and real and imaginary part of the transfer function $H(j\omega)$ is appropriate. However, minimizing the integral-square error may lead to phase errors, and it may be necessary to consider other constraints in order to control phase. Although the integral-square error as a criterion may provide a simplification of analysis and computation, the implications of its use must be carefully considered and understood.

CONSTRAINED AND WEIGHTED APPROXIMATIONS

A few methods which may be used to extend the use of the integral-square error criterion will be discussed briefly. A constrained approximation is one in which the coefficients in series (G-43) are functionally related so that a property of $h^*(t)$ is specified. Mathematically, a constraint may be expressed by

$$k = K\left[h^*(t)\right] = K\left[\sum_{n=1}^{\infty} b_n \phi_n(t)\right] \qquad \text{(G-47)}$$

where K is a functional describing a property of $h^*(t)$, k is a specified value of the property, and

$$h^*(t) = \sum_{n=1}^{N} b_n \phi_n(t) \qquad \text{(G-48)}$$

is the constrained approximation. Examples of possible constraints are:

(a) Having the area under $h^*(t)$ equal unity which is equivalent to normalizing the transfer function, i.e., $H^*(0) = 1$.

$$1 = \int_{-\infty}^{\infty} h^*(t)\,dt = \sum_{n=1}^{N} b_n \int_{-\infty}^{\infty} \phi_n(t)\,dt .$$

(b) Requiring that $H^*(j\omega)$ have a phase shift $\pm 180°$ at $\omega = \infty$, or equivalently, making $h^*(t)$ equal zero at $t = 0$.

$$0 = h^*(0) = \sum_{n=1}^{N} b_n \phi_n(0) .$$

(c) The integral-square value of $h^*(t)$ equals unity

$$1 = \int_{-\infty}^{\infty} h^*(t)^2 dt = \sum_{n=1}^{\infty} b_n^2 .$$

Conditions (a) and (b) are linear in the b_n, while condition (c) is nonlinear and is equivalent to normalizing the average power contained in the impulse approximation.

A weighted approximation is one which attempts to improve the integral-square error. The weighted integral-square error $\mathcal{E}_w$ may be written as

$$\mathcal{E}_w = \int_{-\infty}^{\infty} \left[h(t) - h^*(t)\right]^2 W(t)\, dt \tag{G-49}$$

$$= \int_{-\infty}^{\infty} \left[W^{1/2}(t)\, h(t) - W^{1/2}(t)\, h^*(t)\right]^2 dt . \tag{G-50}$$

Thus, $W^{1/2}(t)\, h(t)$ may be approximated by

$$W^{1/2}(t)\, h^*(t) = \sum_{n=1}^{N} a_n \phi_n(t) . \tag{G-51}$$

From the theory of orthogonality,

$$a_n = \int_{-\infty}^{\infty} W^{1/2}(t)\, h(t)\, \phi_n(t)\, dt \tag{G-52}$$

and the weighted approximation $h^*(t)$ is

$$h^*(t) = W^{-1/2}(t) \sum_{n=1}^{N} a_n \phi_n(t) \tag{G-53}$$

where the functions $\phi_n(t)$ are now orthonormal with respect to the weighting function $W^{1/2}(t)$.

REALIZABILITY OF $h^*(t)$

Realization of a network may be readily accomplished when the transfer function of the network is known as a ratio of polynomials with real coefficients. Thus, if the problem of obtaining the Fourier transform of the impulse response as a ratio of polynomials is solved, then it is possible to complete the realization of the network.

As a result of the requirements for physical realizability of a lumped linear stable system, the impulse response approximation $h^*(t)$ of the system consists of exponential functions and has the form

$$h^*(t) = \sum_{n=1}^{N} R_n e^{s_n t} \tag{G-54}$$

where

1. R_n and s_n may be real or complex,

2. $h^*(t)$ is a real function,

3. s_n has a negative real part.

Equation (G-54) may also be written as

$$h^*(t) = \sum_{i=1}^{k} a_i e^{-b_i t} + \sum_{i=1}^{m} A_i e^{-\alpha_i t} \sin \omega_i t + \sum_{i=1}^{p} B_i e^{-\alpha_i t} \cos \omega_i t . \qquad \text{(G-55)}$$

Hence, only three types of terms can contribute to the overall impulse response $h^*(t)$. These are

$$g_1(t) = ae^{-bt} \qquad \text{(G-56.a)}$$

$$g_2(t) = Ae^{-\alpha t} \sin \omega t \qquad \text{(G-56.b)}$$

$$g_3(t) = Be^{-\alpha t} \cos \omega t . \qquad \text{(G-56.c)}$$

In general, the quantities b, α, and ω are positive real values and independent of time, while the quantities a, A, and B may have any real value and may also be functions of time.

When a transfer function is expressed as a rational fraction, the roots of the polynomial in the denominator are called poles while those of the numerator are referred to as zeros. Every transfer function can be expanded into partial fractions with terms for each pole and a corresponding time function. The impulse response would then be the sum of the time functions associated with the poles. The significance of the poles is that the form of the relating time function is determined by their location in the "$j\omega$" or "s" plane. For example, the poles may be:

1. Real and negative . . .

   The mode of response is a decaying exponential

2. Zero . . .

   The mode of response is a constant

3. Purely imaginary, two roots form conjugate pair . . .

   The pair of modes combines to form a sinusoid

4. Complex with negative real parts, two roots forming a conjugate pair . . .

   The pair of modes combines to form a damped sinusoid

The approximation may be improved for a given number of poles by shifting the zeros relative to each other. A change in the location of the poles alters the quantities in (G-56.a), (G-56.b) and (G-56.c) and changes $h^*(t)$. The zeros of the transfer function contribute to amplitudes and phase angles but do not influence the form of the time function as do the poles.

REALIZABILITY OF SAMPLING METHODS

Early discussions of sampling theorems related the number of discrete values necessary to reproduce a time function. Subsequent discussions of physical elements have shown that the concept of an instantaneous sample is not possible since circuits cannot respond in a nonzero interval of time. Practically, multiplying a signal by a train of impulses actually involves multiplying it by a train of pulses with the duration of each being a finite time, say t_o, and the interval between pulses being the sampling time T.

Figure (G-9) illustrates sampling with samples of nonzero duration. A Fourier analysis of the unit sampling wave $g(t)$ yields a dc term plus harmonics of the sampling frequency. This can be expressed mathematically as

$$g(t) = K + 2K \sum_{n=1}^{\infty} \frac{\sin nK\pi}{nK\pi} \cos \frac{2\pi nt}{T} \tag{G-57}$$

where K is the ratio of the pulse duration to the interval between pulses, i.e., $K = t_o/T$. By passing the product of the signal $f(t)$ and the sampling function $g(t)$ through a low-pass filter, a replica of the signal is obtained. This representation is reduced in magnitude by a factor K (neglecting the delay and any distortion caused by the low-pass filter). Amplifying by $1/K$ will then restore it to its original value. The spectrum of the sampled signal will be the spectrum of the original signal, reduced in magnitude, plus upper and lower sidebands about the sampling frequency f_c and its harmonics.

Another important characteristic of circuit elements is that amplitude and phase response characteristics are not independent. Their relationships may be formulated explicitly by requiring zero response prior to the time that the input is applied. The amplitude and phase characteristics that define an idealized low-pass filter are not realizable in a physical network. However, although an ideal filter is nonrealizable, it may be approximated by physically realizable elements to within a specified accuracy. The closer the approximation, the longer the delay or time of propagation from input to output and the longer the duration of transients in the output in response to frequency components approaching cutoff. In the actual design using sampling principles, the factors which affect the use of bandwidth are:

(1) the tolerance to delay,

(2) the tolerance to the deformation of the sampled wave in the output due to transients, and

(3) the required precision of resolution.

Depending upon the manner and extent to which the highest frequency components of the sampled wave exceed W, $2W$ samples per second may not adequately represent the arbitrary wave being sampled. Practically, there are always limitations of bandwidth and hence there is always an "uncertainty" in the operation on a signal wave by physical elements. By controlling the transmission, at the expense of delay, the uncertainty can be made, in theory, arbitrarily small. It is important to recognize that the uncertainty exists in the absence of other perturbing influences. In most problems it is necessary to consider not only structural components with the associated realizability requirements, but also informational aspects which must necessarily encompass noise and component tolerances. Certain aspects of these interrelationships will be discussed in later sections.

12. TIME-VARYING ELEMENTS

INTRODUCTION

A time-varying element is one where the coefficients of the differential equation describing its behavior are functions of time. If they are independent functions of time, the element is said to be linear. The relation between input and output of a time-varying system can be expressed in a variety of ways other than those based on the use of differential equations. The need for alternative representations is evident, since many time-varying problems such as those involving randomly-varying media and fluctuating targets cannot be characterized by ordinary differential equations of finite order.

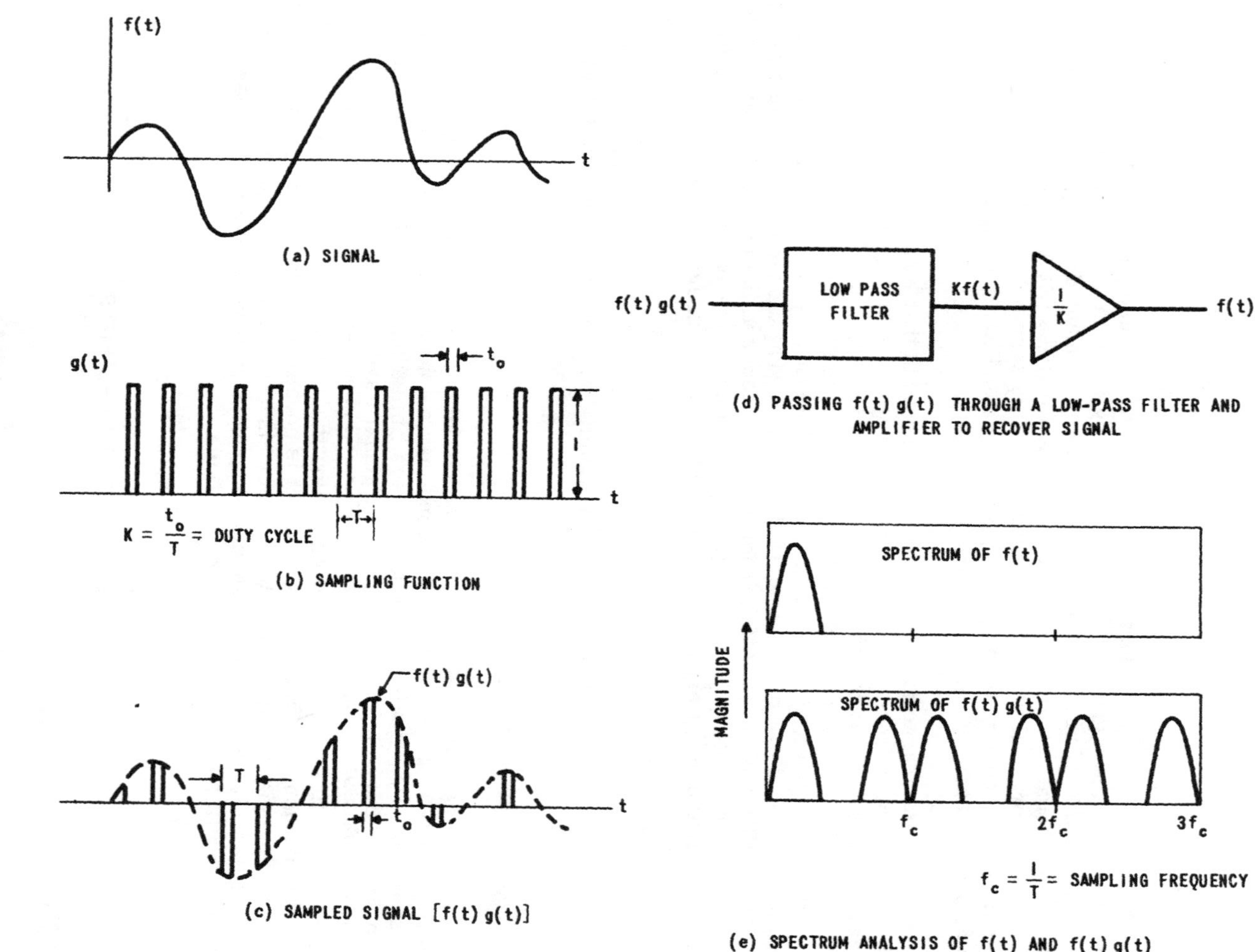

Figure G-9 - Sampling with samples of nonzero duration

A primary reason for considering time-varying elements is that they permit generalization of linear network theory. Time-varying elements are the circuit counterparts to waveforms having time-varying spectra. This suggests that descriptions which were used to characterize time-dependent functions may be applicable for element analysis.

CLASSICAL THEORY OF DIFFERENTIAL EQUATIONS

With the exception of linear equations with variable coefficients which are reducible to those with constant coefficients by a change of variable, there are no general methods for solving such equations of order higher than the first. In general, solutions of differential equations with variable coefficients cannot be expressed in terms of a finite number of elementary functions, and lead to new functions which are defined either by definite integrals or by infinite series, such as Bessel or Legendre functions.

For a time-varying element of input $c(t)$ and output $r(t)$, the homogeneous solution of its linear differential equation of order n,

$$a_o(t)\frac{d^n r(t)}{dt^n} + a_1(t)\frac{d^{n-1} r(t)}{dt^{n-1}} + \cdots + a_{n-1}(t)\frac{dr(t)}{dt} + a_n(t)\, r(t) =$$

$$b_o(t)\frac{d^m c(t)}{dt^m} + b_1(t)\frac{d^{m-1} c(t)}{dt^{m-1}} + \cdots + b_{m-1}(t)\frac{dc(t)}{dt} + b_m(t)\, c(t) \tag{G-58}$$

where the $a_i(t)$ and $b_i(t)$ are continuous single-valued functions of t, possesses (n) linearly independent solutions. If these solutions are $\phi_1(t), \phi_2(t), \ldots, \phi_n(t)$, then the general solution $r_H(t)$ is given by

$$r_H(t) = c_1\phi_1(t) + c_2\phi_2(t) + \cdots + c_n\phi_n(t) \tag{G-59}$$

where the c_i are constants. The simplicity of (G-59) indicates it may often be of practical importance to know whether a given set of functions is linearly independent. The necessary and sufficient condition that a given set of homogeneous solutions $\phi_1(t), \phi_2(t), \ldots, \phi_n(t)$ be linearly independent is that the determinant

$$W \equiv \begin{vmatrix} \phi_1 & \phi_2 & \cdots & \phi_n \\ \phi_1' & \phi_2' & \cdots & \phi_n' \\ \phi_1'' & \phi_2'' & \cdots & \phi_n'' \\ \vdots & \vdots & & \vdots \\ \phi_1^{(n-1)} & \phi_2^{(n-1)} & \cdots & \phi_n^{(n-1)} \end{vmatrix} \neq 0 . \tag{G-60}$$

This is called the Wronskian determinant.

Differential equations of many elements involving time-varying parameters may be solved by direct methods. An example of this are first-order linear equations of the following form,

$$\frac{dr(t)}{dt} + a(t)\cdot r(t) = c(t) \tag{G-61}$$

where $a(t)$ and $c(t)$ are independent functions of time. If the dependent variable $r(t)$ has the initial value,

$$r(0) = 0 \tag{G-62}$$

the general solution of (G-61) is

$$r(t) = \exp\left[-\int a(t)\,dt\right] \cdot \int_0^t \exp\left[a(t)\,dt\right] c(t)\,dt \tag{G-63}$$

using the method of substitutions. Other methods, such as the variation of parameters and those employing differential operators are available.

The physical problem may permit approximations to be obtained readily with sufficient accuracy for practical use. A very useful one is the B.W.K. approximation, used with time-varying elements having parameters which exhibit only small variations about a large average value, and a differential equation of the form:

$$\frac{d^2 r(t)}{dt^2} + a^2(t)\, r(t) = c(t)\,. \tag{G-64}$$

If $a^2(t)$ is a real positive function and satisfies the condition given, a useful approximation to the general solution $r_H(t)$ is

$$r_H(t) = \frac{1}{\sqrt{a(t)}}\left(A \cos\left[q(t)\right] + B \sin\left[q(t)\right]\right) \tag{G-65}$$

where A and B are arbitrary constants, and $q(t)$ is given by

$$q(t) = \int a(t)\,dt\,. \tag{G-66}$$

It is particularly good if the variations of $a(t)$ are such that

$$\left|a^2(t)\right| \gg \left|\frac{a''(t)}{2a(t)} - \frac{3}{4}\left(\frac{a'(t)}{a(t)}\right)^2\right| \tag{G-67}$$

in the range of (t) under consideration. Other approximations exist, depending on the type of elements involved and the temporal range of their parameters.

TRANSFER FUNCTION

It was shown in section (G-3) that a linear time-invariant element can be usefully characterized through knowledge of its transfer function $H(j\omega)$. Transfer functions can be similarly applied to linear time-varying elements. Since they are functions of both time and frequency, they are designated by $H(j\omega; t)$.

By introducing the Heaviside operator $p = d/dt$, the problem of characterizing the element, that is, the solving of the differential equation (G-58), may be simplified to determining the ratio of input to output, $H(p; t)$,

$$H(p;\, t) = \frac{r(t)}{c(t)} = \frac{b_o(t)p^m + b_1(t)p^{m-1} + \cdots + b_m(t)}{a_o(t)p^n + a_1(t)p^{n-1} + \cdots + a_n(t)}\,. \tag{G-68}$$

The transfer function $H(j\omega; t)$ of a time-varying element N is defined as

$$H(j\omega; t) = \frac{\text{response of N to } e^{j\omega t}}{e^{j\omega t}}\,. \tag{G-69}$$

Thus, when the input is $e^{j\omega t}$, (G-68) becomes

$$H(j\omega;t) = \left.\frac{r(t)}{c(t)}\right|_{c(t)=e^{j\omega t}} = \frac{b_o(t)(j\omega)^m + b_1(t)(j\omega)^{m-1} + \cdots + b_m(t)}{a_o(t)(j\omega)^n + a_1(t)(j\omega)^{n-1} + \cdots + a_n(t)} \qquad \text{(G-70)}$$

The time-varying frequency-response function $H(j\omega;t)$ constitutes a natural generalization of $H(j\omega)$. If the element is initially at rest and the Fourier transform of the input $c(t)$ is denoted by $C(j\omega)$ the expression for the output at time t, $r(t)$, in terms of $H(j\omega;t)$ and $C(j\omega)$ is

$$r(t) = \frac{1}{2\pi}\int_{-\infty}^{\infty} H(j\omega;t)\, C(j\omega)\, e^{j\omega t}\, d\omega . \qquad \text{(G-71)}$$

Figure (G-10) compares fixed and variable elements for a sinusoidal input.

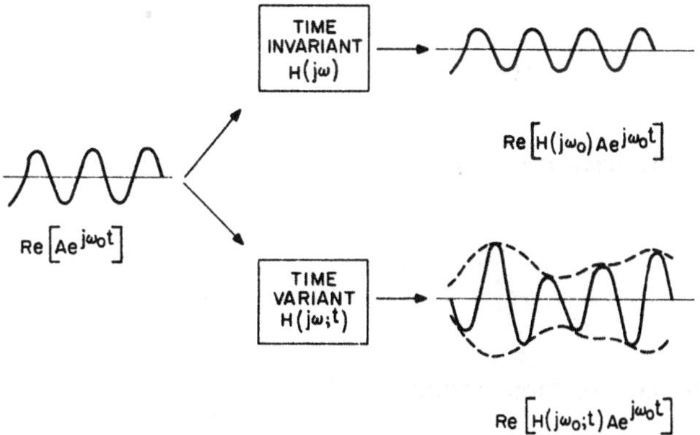

Figure G-10 - Comparison of time-invariant and time-variant elements for a sinusoidal input

When the spectrum varies with time, more weight must be given to values ot the time-function occurring at certain times than at others. This implies a type of modulation. In general, functions with time-varying spectra may be analyzed by considering two types of variations, one of which involves rapid fluctuations of the functions, the other, slow changes.

IMPULSE RESPONSE

A time-dependent transfer function can be associated with a time function which has some characteristic of it varying with time, indicative of the behavior of the element. This is the impulse response of the network and its use retains the advantages discussed previously for time-invariant elements.

If the input $c(t)$ is the impulse function $\delta(t-T)$, then the output $r(t)$ is denoted as $h(t;T)$ and is called the impulse response of the element,

$$h(t;T) = H(p;t)\ \delta(t-T) . \qquad \text{(G-72)}$$

As for a fixed network, the impulse response and transfer function are conjugate Fourier transforms. The Fourier transform of (G-72) with respect to T, considering t as a parameter, is

$$\mathcal{F}\{h(t;T)\} = H(-j\omega;t)\, e^{-j\omega t} . \qquad \text{(G-73)}$$

Taking the inverse Fourier transform and replacing ω by $-\omega$,

$$h(t;\, T) = \frac{1}{2\pi} \int_{-\infty}^{\infty} H(j\omega;t)\, e^{j\omega(t-T)}\, d\omega\,. \qquad \text{(G-74)}$$

Transforming both sides of (G-74) with respect to T,

$$H(j\omega;\, t)\, e^{j\omega t} = \int_{-\infty}^{\infty} h(t;\, T)\, e^{-j\omega T}\, dT\,. \qquad \text{(G-75)}$$

Equations (G-74) and (G-75) establish the conjugate behavior of $h(t;\, T)$ and $H(j\omega;\, T)$. Since $H(j\omega;\, t)$ can describe a network completely and is related to $h(t;\, T)$ by (G-75), then $h(t;\, T)$ also completely describes the network.

In general, the output $r(t)$ is related to the input $c(t)$ through the impulse response, i.e.,

$$r(t) = \int_{-\infty}^{\infty} h(t;\, T)\; c(T)\, dT\,. \qquad \text{(G-76)}$$

For a physically realizable network, $h(t;\, T) = 0$ for $T > t$, and $c(t) = 0$ for $T < 0$. (G-76) then reduces to

$$r(t) = \int_{0}^{t} h(t;\, T)\; c(T)\, dT \qquad \text{(G-77)}$$

and is referred to as the convolution integral.

GREEN's FUNCTION

A method which is useful in describing time-varying elements involves the use of Green's function. Green's function is used to characterize the network's impulse response and the differential equation which describes its behavior. (G-58) may be written as

$$L\, r(t) = y(t) \qquad \text{(G-78)}$$

where L is the linear differential operator

$$L = a_o(t) \frac{d^n}{dt^n} + a_1(t) \frac{d^{n-1}}{dt^{n-1}} + \cdots + a_n(t) \qquad \text{(G-79)}$$

and $y(t)$ is a known function of the input, namely

$$y(t) = b_o(t) \frac{d^m c(t)}{dt^m} + b_1(t) \frac{d^{m-1} c(t)}{dt^{m-1}} + \cdots + b_m(t)\; c(t). \qquad \text{(G-80)}$$

Subject to the boundary conditions

$$r^{(\alpha)}(0) = 0 \qquad \text{for} \quad \alpha = 0, 1, \ldots, n-1 \qquad \text{(G-81)}$$

the solution of the differential equation (G-78) becomes

$$r(t) = \int_{0}^{t} G_1(t;\, x)\; y(x)\, dx \qquad \text{(G-82)}$$

where $G_1(t;\, x)$ is the one-sided Green's function. The latter is defined as

$$G_1(t;\, x) = \frac{(-1)^{n-1}}{b_o(x)\; W(x)} \begin{vmatrix} \phi_1(t) & \phi_2(t) & \cdots & \phi_n(t) \\ \phi_1(x) & \phi_2(x) & & \\ \phi_1'(x) & \phi_2'(x) & & \\ \cdots & \cdots & \cdots & \cdots \\ \phi_1^{(n-2)}(x) & \phi_2^{(n-2)}(x) & & \phi_n^{(n-2)}(x) \end{vmatrix} \tag{G-83}$$

where $W(x)$ is the Wronskian of the linearly independent solutions $\{\phi_1(t),\, \phi_2(t),\, \ldots,\, \phi_n(t)\}$ of $L\, r(t) = 0$. Since the upper limit in (G-82) is variable, the integral equation is of the Volterra type.

If N is the linear differential operator

$$N = b_o(t)p^m + b_1(t)p^{m-1} + \cdots + b_m(t) \tag{G-84}$$

then

$$Nc(t) = y(t) \tag{G-85}$$

and (G-82) becomes

$$r(t) = \int_0^t G_1(t;\, x)\, Nc(x)\, dx. \tag{G-86}$$

Thus, Green's function completely characterizes the network for a given input, the output being uniquely determined. Equations (G-82) and (G-86) are similar in form to (G-77). For $N = 1$, the relation between the impulsive response and the one-sided Green's function of a linear network is

$$\left.\begin{aligned} h(t;\, T) &= G_1(t;\, T) \qquad t > T \\ &= 0 \qquad\qquad\quad\; t < T \end{aligned}\right\}. \tag{G-87}$$

If N cannot be expressed as (G-84), then G_1 is not the same as h.

The advantage of Green's function in investigating linear systems is ascribed to the function's properties and the ease with which physical interpretations may be made. Some of the properties are:

(a) $\left.\dfrac{\partial^\alpha}{\partial t^\alpha} G_1(t;\, x)\right|_{t=x} = 0 \quad$ for $\alpha = 0, 1, \ldots, n-2$

(b) $\left.\dfrac{\partial^{n-1}}{\partial t^{n-1}} G_1(t;\, x)\right|_{t=x} = \dfrac{1}{b_o(x)}$

(c) $L_t\, G_1(t;\, x) = 0$

(d) $G_1(t;\, x)$ is unique

(e) Given $G_1(t;\, x)$, a set of solutions of $L\, r(t) = 0$ can be determined explicitly.

[L_t implies that the differential operator L operates on functions of t.]

For the general boundary conditions of the form

$$U_\alpha(r) = \sum_{\beta=1}^{n} A_{\alpha\beta} r^{(\beta-1)}(a) + \sum_{\beta=1}^{n} B_{\alpha\beta} r^{(\beta-1)}(b) = 0 \qquad \alpha = 1, 2, \ldots, n \tag{G-88}$$

where a and b are two distinct time instants, $b > a$, the solution of (G-78) is

$$r(t) = \int_a^b G_2(t;\, x)\, y(x)\, dx \tag{G-89}$$

where $G_2(t;\, x)$ is the two-sided Green's function. Since both limits are fixed, (G-89) is a Fredholm integral equation.

From section F-4, the Volterra and Fredholm integral equations can be reduced to algebraic form through the use of integral transforms, the Laplace and Mellin transforms, respectively. Thus, integral transforms can be considerably useful in solving problems concerning time-varying elements.

INTEGRAL TRANSFORMS

The use of integral transforms to solve differential equations of specific time-varying elements has already been discussed in section F-4. To review briefly, the application of an integral transformation method to linear networks is based upon first resolving the solutions of the differential equations into an integration (or summation) of elementary functions $k(t;\, s)$ where (s) may be considered a complex parameter.

For fixed, that is, invariant, elements, resolving the solution $r(t)$ consists essentially in expressing it in the following form

$$r(t) = \int_C k(t;\, s)\, R(s)\, ds \tag{G-90}$$

where C is the contour in the s-plane (generally a straight line parallel to either the imaginary or real axis) and $R(s)ds$ is a weighting factor which provides a measure of the content in $r(t)$ of those components $k(t;\, s)$ in which the parameter lies between s and $s + ds$. $R(s)$ is called the spectral function of $r(t)$ relative to $k(t;\, s)$. Due to the linear nature of (G-90), the expression for $R(s)$ in terms of $r(t)$ is of the general form

$$R(s) = \int_{-\infty}^{\infty} K(s;\, t)\, r(t)\, dt \tag{G-91}$$

where $K(s;\, t)$ denotes the inverse of $k(t;\, s)$. To complete the uniqueness of a spectral description, the desired relation between $k(t;\, s)$ and its inverse is

$$\int_C k(t;\, s)\, K(s;\, \xi)\, ds = \delta(t - \xi) \tag{G-92}$$

where $\delta(t - \xi)$ is the delayed unit impulse.

For time-varying elements, (G-91) may be modified as

$$R(s;\, t) = \int_{-\infty}^{t} K(s;\, \lambda)\, r(\lambda)\, d\lambda\, . \tag{G-93}$$

This is a time-variable transform which maps a function in the time or t-domain into a generalized $s;\, t$ domain, where t behaves like a parameter. Thus, (G-93) is a "running transform," similar to that used to describe the instantaneous power spectra (B-16) for various waveforms having time-varying spectra. The relation between the elementary function and its inverse is as defined by (G-92). The inverse transform is then

$$r(t) = \int_C k(t;\, s)\, R(s;\, t)\, ds\,. \tag{G-94}$$

The set of functions $k(t;\, s)$ constitutes a coordinate system in some vector space. Transformation from $r(t)$ to $R(s;\, t)$ implies a decomposition of $r(t)$ along the axis of the coordinate system which is dependent of time.

The time-variable transform will inherently contain the properties of the single-variable transform. However, since it involves an integral over finite limits, it has additional properties. A theorem resembling Parseval's may be applied when $K(s;\, t)$ is described over the contour C by

$$K(s;\, t) = f(t)\, k^*(t;\, s) \tag{G-95}$$

where $f(t)$ is a function of t for all s over C and the asterisk denotes the complex conjugate. If $r(\xi)$ is bounded for $0 \leq \xi \leq t$ the theorem then states, that:

$$\int_C |R(s;\, t)|^2\, ds = \int_0^t f^*(\xi)\, r^2(\xi)\, d\xi\,. \tag{G-96}$$

Other integral transforms exist for the solution of physical problems. The type of transformation performed — specifically, the kernel function used — depends on the linear system and associated initial conditions. A method of developing transforms for any linear system is to introduce boundary conditions in the way that the Laplace transform does. As an example, consider the following differential equation describing a time-varying element

$$a(t)\, r''(t) + b(t)\, r'(t) + d^2 r(t) = c(t) \tag{G-97}$$

where $a(t)$ and $b(t)$ are functions of time and d^2 is a constant. It is desired to reduce this to an algebraic equation in the transform domain, such as

$$q(s)\, R(s) + d^2 R(s) = C(s) + \left[\text{terms involving initial conditions}\right], \tag{G-98}$$

when the integral transformation of the form

$$R(s) = \int_0^\infty K(s;\, t)\, r(t)\, dt \tag{G-99}$$

is applied, with $q(s)$ being an arbitrary function of the transform variable.

The kernel, $K(s;\, t)$, is obtained by applying (G-99) to (G-97) and integrating by parts. If a function of time $g(t)$ is included in the kernel to make the linear differential operator of (G-97) self-adjoint, the kernel becomes

$$K(s;\, t) \equiv g(t)\, \ell(s;\, t) \tag{G-100}$$

where $g(t)$ is given by

$$g(t) = \exp\left[\int \frac{b(t) - a'(t)}{a(t)}\, dt\right] \tag{G-101}$$

For these conditions, $\ell(s;\, t)$ must satisfy

$$a(t)\, \ell_{tt}(s;\, t) + b(t)\, \ell_t(s;\, t) - q(s)\, \ell(s;\, t) = 0 \tag{G-102}$$

where ℓ_t and ℓ_{tt} refer to the first and second derivative with respect to time, respectively. (G-102) shows that the appropriate transform depends on a knowledge of the solution of the homogeneous form of the system equation (G-97).

The transform method is most useful when the system under consideration is subjected to various excitations. It is important to note that the philosophical difference between the finite upper limit of (G-93) and the infinite upper limit of (G-99) is that the former produces a transform which varies with time, its value at any one instant being independent of future values of the response. The latter produces a transform, such as the Laplace and Mellin transforms, which depends on the entire history of the waveform.

RANDOMLY-VARYING ELEMENTS

Up to this stage, only deterministic time-varying elements have been discussed. They have been represented by differential equations with variable coefficients whose values may be predicted with probability one at future instants of time. Their descriptions are special cases of a more general approach which considers statistical characteristics.

Randomly-varying elements, those whose parameters vary randomly with time, are becoming of increasing importance and their analysis permits a more general classification of elements. Consider a nonrandom input $c(t)$ applied to a randomly-varying element whose behavior may be expressed by

$$a_o(t)\,\frac{d^n r(t)}{dt^n} + \cdots + a_n r(t) = c(t)\,. \tag{G-103}$$

If $a_i(t)$ varies as

$$a_i(t) = \overline{a_i(t)} + \epsilon_i(t) \tag{G-104}$$

with $\overline{a_i(t)}$ being the expected value of $a_i(t)$ and $e_i(t)$ is small compared with $\overline{a_i(t)}$, then (G-103) can be solved by perturbation techniques, with $r(t)$ being the sum of a nonrandom term and a random term ascribed to $\epsilon(t)$.

If characterization by a differential equation is impractical, correlation and spectral analysis may be used. Similar to the treatment applied to descriptions of random waveforms, it is convenient to assume stationarity in analysis. Let $[u(t)]$ and $[v(t)]$ represent two independent stationary processes which, when applied separately to a randomly-varying element, result in processes having autocorrelation functions $\psi_u(\tau)$ and $\psi_v(\tau)$, respectively. It is assumed the inputs are independent of the random processes governing the behavior of the element. If by applying process $[\alpha u(t) + \beta v(t)]$ to the input, where α and β are arbitrary real constants, a random process is produced whose correlation function $\psi_{\alpha u+\beta v}(\tau)$ is given by:

$$\psi_{\alpha u+\beta v}(\tau) = \alpha^2\,\psi_u(\tau) + \beta^2\,\psi_v(\tau) \tag{G-105}$$

for all $\alpha, \beta, [u]$ and $[v]$, then the element is said to be linear. Linearity, here, implies the superposition property for correlation functions. Any stationary linear element will have this property. (G-105) can be used as a basis for determining whether an element is linear by observing the input and output over periods of time sufficiently long to enable obtaining accurate estimates of the correlation functions involved in (G-105).

The correlation function of a stationary randomly-varying element is defined as

$$\psi(j\omega;\,\tau) = E\left\{\left[H(j\omega;\,t)\;H(-j\omega;\,t+\tau)\right]\right\} \tag{G-106}$$

where $H(j\omega;\,t)$ is the time-dependent transfer function given by (G-70) together with (G-58). For each real ω, $[H(j\omega;\,t)]$ is a stationary random process. If the input is a stationary random process $[c(t)]$, independent of $[H(j\omega;\,t)]$, then the correlation function of the output process $[r(t)]$ is

$$\psi_r(\tau) = \frac{1}{2\pi} \int_{-\infty}^{\infty} \psi(j\omega; \tau)\ \mathcal{F}\left\{\psi_c(\tau)\right\} e^{j\omega\tau}\, d\omega \qquad \text{(G-107)}$$

where $\mathcal{F}\{\psi_c(\tau)\}$ is the Fourier transform of the input correlation function, $\psi_c(\tau)$. This relation is of the same form as that expressing the output of a time-varying network with transfer function $\psi(j\omega; \tau)$, with the input being $\psi_c(\tau)$. This can be seen by comparing (G-107) with (G-71).

13. CONCLUSION

A number of methods have been reviewed, interrelating various descriptions of linear elements. An attempt has been made to stress the basic similarities of these methods with techniques previously discussed for representing structural detail of functions. Functions discussed earlier included periodic, transient, and random processes which were indicated as being of importance for representing temporal or spatial structure. The basic philosophy involved the concept that the major purposes of representing and transforming structure are as simplification and matching operations. The choice of a particular method is consequently dependent not only on the nature of the function but also on the use which is to be made of the representation. Typical uses with which the discussions were concerned included improving visualization, understanding, and computation, and facilitating physical realization. Since simplification is a highly subjective concept, it is necessary to include additionally, for most applications, quantitative measures of "completeness." Characteristics of the integral-square error as a criterion were discussed together with the use of Fourier transform and related methods applied to time-invariant structures. Sampling, and correlation and spectral analysis were outlined, along with additional descriptions which were required particularly when boundary conditions were imposed simultaneously in conjugate domains.

Analogous relationships were seen to occur when the representations of linear circuit elements were reviewed. Just as the representations of functions were characterized by a wide range of methods and techniques, the analysis of physical elements may also be made in terms of differential-integral equations, transfer functions, and impulse response. These methods may be extended to include time-varying structures. Perhaps the most important single concept is that the number and type of structural components to be used — (whether the problem involves "signals," elements, or the relationships between signals and elements) — are not to be regarded as intrinsic properties, but as convenient reference elements which are dependent on the mode of representation and the nature of the problem.

REFERENCES

G-1. E. Weber, "Linear Transient Analysis," Vol. I, Chap. 2, John Wiley and Sons, Inc., 1957

G-2. W. B. Davenport, Jr. and W. L. Root, "An Introduction to the Theory of Random Signals and Noise," Chap. 9, McGraw-Hill Book Company, Inc., 1958

G-3. S. Goldman, "Information Theory," pp. 219-229, Prentice-Hall, Inc., 1953

G-4. E. G. Gilbert, "Linear System Approximation by Differential Analyzer Simulation of Orthonormal Approximation Functions," I.R.E. Trans., Vol. EC-8, No. 2, pp. 204-209, June 1959

G-5. J. D. Brule', "Improving the Approximation to a Prescribed Time Response," I.R.E. Trans, Vol. CT-6, No. 4, pp. 355-361, December 1959

G-6. J. D. Brule', "Time-Response Characteristics of a System as Determined by its Transfer Function," I.R.E. Trans., Vol. CT-6, No. 2, pp. 163-170, June 1959

G-7. A. H. Zemanian, "Network Realizability in the Time Domain," I.R.E. Trans., Vol. CT-6, No. 3, pp. 288-291, September 1959

G-8. G. Raisbeck, "A Definition of Passive Linear Networks in Terms of Time and Energy," J. Appl. Phys., Vol. 25, pp. 1510-1514, December 1954

G-9. H. M. James, N. B. Nichols, and R. S. Phillips, "Theory of Servomechanisms," Chap. 2, McGraw-Hill Book Company, Inc., 1947

G-10. E. S. Kuh and D. O. Pederson, "Principles of Circuit Synthesis," Chap. 2, McGraw-Hill Book Company, Inc., 1959

G-11. E. A. Guillemin, "Introductory Circuit Theory," John Wiley and Sons, Inc., New York, 1953

G-12. R. M. Fano, "Short-Time Autocorrelation Functions and Power Spectra," J. Acoust. Soc. Am., Vol. 22, No. 5, pp. 546-550, Sept. 1950

G-13. H. S. Black, "Modulation Theory," Chap. 4, D. Van Nostrand Company, Inc., N. J.

G-14. L. A. Pipes, "Four Methods for the Analysis of Time-Variable Circuits," I.R.E. Trans., Vol. CT-2, No. 1, pp. 1-12, March 1955

G-15. J. Brodin, "Analysis of Time-Dependent Linear Networks," I.R.E. Trans., Vol. CT-2, No. 1, pp. 12-16, March 1955

G-16. A. A. Gerlach, "A Time-Variable Transform and Its Application to Spectral Analysis," I.R.E. Trans., Vol. CT-2, No. 1, pp. 22-25, March 1955

G-17. K. S. Miller, "Properties of Impulsive Responses and Green's Functions," I.R.E. Trans., Vol. CT-2, No. 1, pp. 26-31, March 1955

G-18. J. A. Aseltine, "A Transform Method for Linear Time-Varying Systems," J. Appl. Phys., Vol. 25, No. 6, pp. 761-764, June 1954

G-19. L. A. Zadeh, "A General Theory of Linear Signal Transmission Systems," J. of Franklin Institute, pp. 293-312, April 1952

G-20. L. A. Zadeh, "Frequency Analysis of Variable Networks," Proc. I.R.E., Vol. 38, pp. 291-299, March 1950

G-21. L. A. Zadeh, "Time-Varying Networks, I," Proc. I.R.E., Vol. 49, pp. 1488-1503, Oct. 1961

G-22. L. A. Zadeh, "Correlation Functions and Power Spectra in Variable Networks," Proc. I.R.E., Vol. 38, pp. 1342-1345, Nov. 1950

G-23. E. C. Ho and H. Davis, "Generalized Operational Calculus for Time-Varying Networks," Dept. of Engrg., University of California, Los Angeles, Report No. 54-71, July 1954

G-24. J. M. Manley, "Some Properties of Time-Varying Networks," I.R.E. Trans., Vol. CT-7, pp. 69-78, August 1960

G-25. S. Darlington, "An Introduction to Time-Variable Networks," Proc. Symp. on Circuit Analysis, University of Illinois, Urbana, 1955

G-26. B. E. Keiser, "The Linear, Input-Controlled, Variable-Pass Network," I.R.E. Trans., Vol. IT-1, No. 1, pp. 34-39, March 1955

G-27. M. C. Herrero, "Resonance Phenomena in Time-Varying Circuits," I.R.E. Trans., Vol. CT-2, No. 1, pp. 35-41, March 1955

G-28. J. R. Carson and T. C. Fry, "Variable Electric Circuit Theory with Application to the Theory of Frequency Modulation," Bell Sys. Tech. Jour., Vol. 16, pp. 513-540, 1937

G-29. C. H. Page, "Physical Mathematics," Chap. 12-15, D. Van Nostrand Co., Inc., Princeton, N. J., 1955

G-30. W. M. Siebert, "Signals in Linear Time-Invariant Systems," Chap. 3, Lectures on Communication System Theory, McGraw-Hill, N. Y., 1961

H. LINEAR ANALYSIS OF SPATIAL ELEMENTS

1. INTRODUCTION

A number of analogies exist between circuit and spatial elements when linear analysis is applicable in describing behavior of elements. It is important to recognize, however, that the "analogy" may consist solely of a common mathematical formulation, and that the full utility of analogues can only be established by considering the physical nature of the elements' excitation, the environment in which it is to function, and the use which is to be made of the descriptions. Descriptions and their transformations assume practical importance when they serve to portray concisely and completely element behavior and reduce some of the difficulties associated with improved understanding, computing or physical realization. Development of analogies is important since design details and computations made in one area may be used in other areas provided that correct analysis of the physical process has been made. In addition to analytical and physical analogues which exist between circuit and spatial elements, properties of acoustical radiation resemble electromagnetic fields, and as a result characteristics such as the directivity patterns of acoustical transducers may be derived from microwave antenna or optical element configurations, and in some cases acoustical structures may be used for electromagnetic problems. When structural detail of the radiated field is of primary concern, it is convenient to confine the analysis to transmitting sources since directivity patterns are identical for transmission and reception when linear, reciprocal elements are involved.

Methods for describing basic properties of spatial elements may use space, or space-frequency as variables. The freedom of selection corresponds to the choice of time or frequency in circuit problems. Similarly, correlation and spectral analyses, and statistical methods may be employed. As was indicated in earlier discussions, the use of statistical methods may be required in order to reduce dimensionality, or may be required because the only available information is statistical in nature, and in many problems combinations of deterministic and probabilistic descriptions are required.

2. LINEAR SUPERPOSITION

A spatial element is said to be linear if it obeys the law of linear superposition. The practical result is that the directive properties of a spatial element such as an acoustic line transducer may be determined by examining the behavior of a number of discrete receiving points spaced along the line. In the case of longitudinal waves such as sound, a point source would consist of a sphere which is small in comparison with the wavelength emitted so as to radiate a spherical wave. Surfaces of constant amplitude are spheres concentric with the source. For electromagnetic waves which are transverse, the electric field, the magnetic field, and the direction of propagation are perpendicular to each other. A point source is represented by a dipole which may be considered to be a short wire carrying the current, with the length of the wire being small in comparison with the wavelength. An expression may be derived for the response of a single point source. The amplitude and phase of this response are expressed as functions of the position of the element with respect to a transmitting point. The total response is obtained by combining the elementary responses for the spatial device through an integration which relates the spatial configuration and hence, will be a function of the dimensions and bearing of the incident radiation.

Figure (H-1) shows a comparison of the directivity patterns of a discrete linear array having elements spaced at intervals of $\lambda/2$ and a continuous, uniform linear transducer. Each have a total length of 5λ and the relative acoustic pressure is specified by the ratio of the voltage developed by the acoustic energy of given intensity, at a given bearing, to that developed

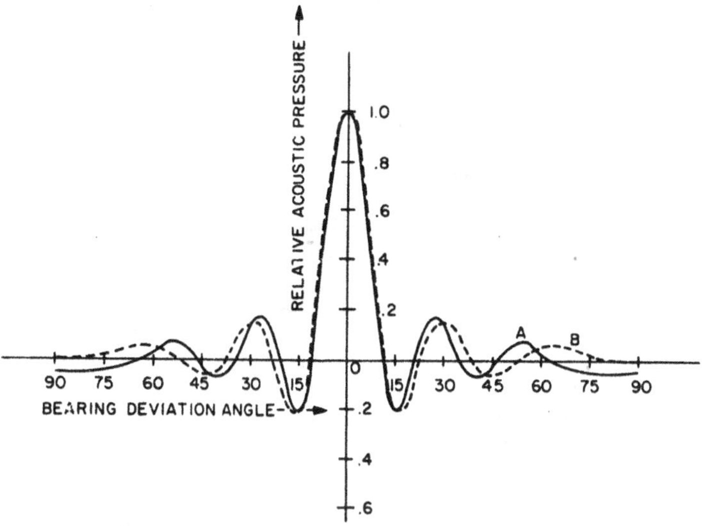

Figure H-1 - A comparison of the directivity patterns of (A) a discrete linear array having elements spaced at half wave lengths and (B) a continuous, uniform line transducer

by the acoustic energy of the same intensity arriving along the axis normal to the array. A line transducer is equivalent to a discrete array having an infinite number of elements. It is seen that there is little improvement in directionality to warrant providing a discrete array with more elements than are sufficient to give a half-wave spacing. The height of the secondary lobes may be slightly increased but the width of the major lobe is practically unchanged.

When the responses of discrete receiving elements are known, the response of an arbitrary configuration can be derived by using the principle of pattern multiplication. The radiation pattern of an array of spatial elements, each of which has the same pattern with the same orientation in space, may be found by (1) replacing each of the elements by an omnidirectional element at the same point and with the same amplitude and phase of excitation, (2) determining the array pattern of the resulting array of omnidirectional elements, and (3) multiplying the array pattern by the radiation pattern of the individual elements of the original array.

A linear array of N equally-spaced elements has N degrees of freedom since it is possible to establish N coefficients of the Fourier series for the total far-zone pattern. As a result, N linearly independent aperture distributions are available, and consequently, N points may be assigned to the radiation pattern. In the general case, where the elements are arbitrarily distributed, each element will have an added degree of freedom, namely, its position along the axis of the array. Therefore, the array with arbitrarily distributed elements needs, in general, fewer elements. This is analogous to nonuniform temporal sampling of a signal.

3. POINT SOURCE RESPONSE; CONVOLUTION THEOREM

The analogy to the impulse response in the time domain is the response of a spatial element to a point source. Such a source radiates uniformly in all directions and is the building block needed for linear superposition.

In two dimension if $h(x,y)$ is the point source response, then subject to linear superposition, the total response $r(x,y)$ for an extended source distribution $s(u,v)$ is given by

$$r(x,y) = \int_{-\infty}^{\infty} \int_{-\infty}^{\infty} h(x-u,\ y-v)\ s(u,v)dudv \tag{H-1}$$

where (x,y) and (u,v) are the spatial coordinates for the output of the spatial element and the source, respectively. Hence, analogous to a temporal system, the output of a spatial element is the convolution of its point source response and input. This is the spatial convolution theorem. Although Eq. (H-1) can be readily formulated, it cannot always be integrated in closed form and either numerical analysis or transfer function descriptions may be required.

The directional discrimination of a receiver depends on the response to sources outside the major lobe. Boundaries imposed on spatial extent produce sidelobes in the response and affect the interpretation of the various spatial descriptions. For example, a limitation to perfect transmission in a coherent system is the reciprocal relationship between the wavelength of the radiation and the spacing between detail in the object (or source). This indicates that information cannot be transmitted in closer detail than the wavelength of the incident radiation.

The plane wave response pattern of a receiving antenna is obtained when the radiator is a point source at a sufficient distance such that an increase in the distance will produce no detectable change in the pattern. If the source subtends an appreciable angle, the response pattern will be modified. This is shown in Figure (H-2) where the pattern of a receiving antenna is compared to the pattern observed when the point source is replaced by an extended source, at the same distance.

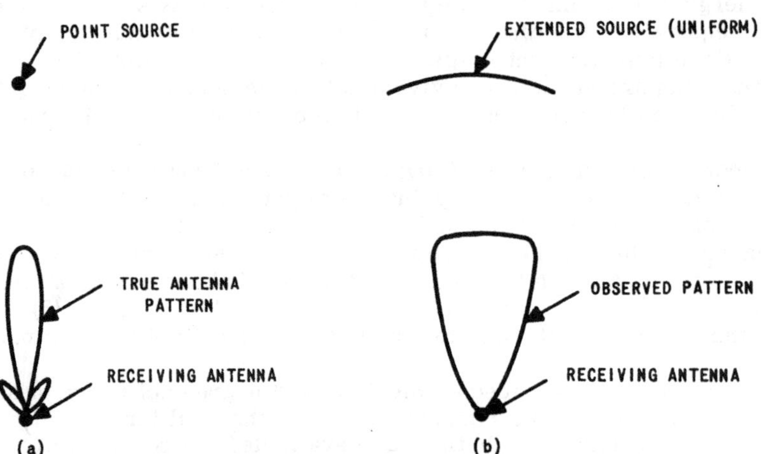

Figure H-2 - Antenna pattern for (a) a point source and (b) an extended source

The effect of the source distribution on the observed power pattern $G(\phi_o)$ may be given as

$$G(\phi_o) = \frac{1}{A}\int F(\phi + \phi_o)\ f(\phi)\ d\phi \tag{H-2}$$

where

$G(\phi_o)$ = observed or resultant pattern,

$F(\phi + \phi_o)$ = true antenna pattern (as measured with a point source),

$A = \int f(\phi)\ d\phi$ = effective angle subtended by source (total power flux of source),

$f(\phi)$ = source distribution.

All patterns in the above equation are proportional to power and are shown in Figure (H-3) where the main lobe of the antenna is displaced from the center line of the source by an angle ϕ_o and the over-all source extent is 2α. For a point source, the source pattern in Figure (H-3) reduces to an impulse at $\phi = 0$ $(\alpha = 0)$ and $G(\phi) = F(\phi)$. Thus, for a point source, the observed pattern is identical with the true pattern.

Figure (H-4) illustrates the case of an extended source that is much broader than the antenna pattern, and with the source being represented by a step function equal to unity between $+\alpha$ and $-\alpha$ and zero elsewhere. In the range of ϕ_o between $\alpha - \beta$ and $-(\alpha - \beta)$ the observed distribution is constant but reduced by a factor B/A, where (B) is the area under the antenna pattern and (A) is the area under the source pattern.

In general, the pattern $F(\phi + \phi_o)$ and the observed pattern $G(\phi_o)$ are known while the source distribution $f(\phi)$ is unknown. The latter can be determined by assuming various source distributions and calculating the corresponding distributions, $G(\phi_o)$. If the calculated $G(\phi_o)$ distribution agrees with the actual observed distribution, then the assumed source distribution $f(\phi)$ represents the true source distribution or its equivalent. A more direct method is to expand the distributions into Fourier series and relate the corresponding coefficients as dictated by Eq. (H-2).

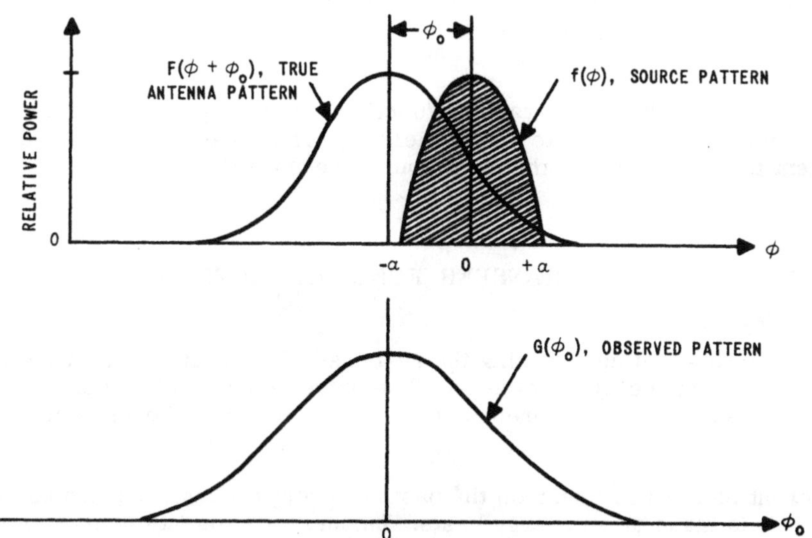

Figure H-3 - Antenna pattern, source pattern, and resultant or observed pattern

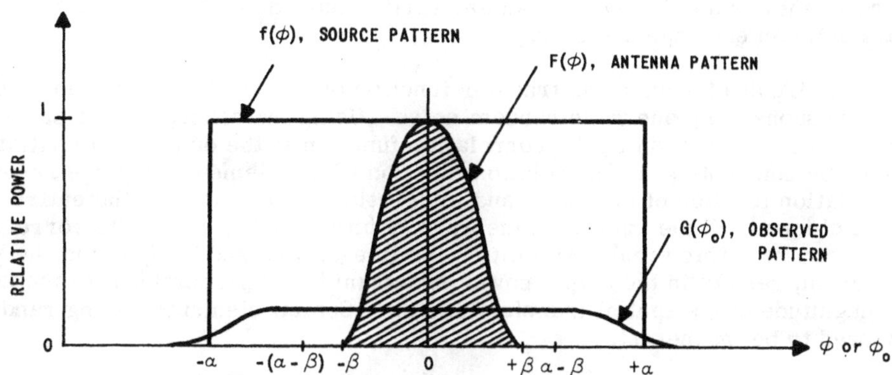

Figure H-4 - Case of source pattern that is much wider than antenna beamwidth

4. TRANSFER FUNCTION

From the properties of Fourier transforms, the space-frequency spectrum of an aperture distribution modified by any number of successive linear operations is the product of the space-frequency spectrum of the original distribution, and the spatial spectra of the several linear operations. Therefore, an analysis performed in the space-frequency domain replaces successive integrations by successive multiplications. The space-frequency spectrum of the response of a spatial element may be expressed as

R	=	(S)	(H)	
response spatial spectrum		source spatial spectrum	transfer function of spatial element	(H-3)

H is the Fourier transform of the point source response whose coordinates are spatial frequencies having dimensions of reciprocal length. For an n-dimensional distribution this is written as

$$H(s_1, s_2, \ldots, s_n) = \underbrace{\int_{-\infty}^{\infty} \cdots \int_{-\infty}^{\infty}}_{n\text{-fold}} h(x_1, x_2, \ldots, x_n) e^{j2\pi(x_1 s_1 + \ldots + x_n s_n)} dx_1 \ldots dx_n \qquad \text{(H-4)}$$

where $s_j = -\sin\theta_j/\lambda$. The space transfer function completely characterizes a spatial element. It is primarily a steady-state or far-field description and, as indicated above, is often more convenient to work with than the point source response.

5. TRANSFER FUNCTION METHOD

The general problem of determining the response of a spatial element requires knowledge of the degree of coherence of the excitation. The extreme cases of completely coherent and incoherent elements are readily represented; however, partial coherence characterizes real elements.

An incoherent element is linear on the basis of energy, permitting the use of Fourier analysis only if it is made on an energy basis. Incoherent elements behave as low-pass filters since they deal with the addition of nonnegative intensity variations. The inherent flexibility associated with spatial-frequency operations is lost when incoherent elements comprise the system. A coherent element is linear in amplitude and phase. The Fourier components that make up the response may be controlled by using the proper weighting with the element. In reception, a coherent element behaves as an amplitude-phase detector whereas an incoherent element is basically an envelope detector.

Analogous to circuit elements, the transfer function of a spatial element can be determined from its response to plane wave random noise. Using the concept of the space correlation function developed in section E, the correlation function at the output of a spatial element may be obtained by convoluting the correlation function of the elements impulse response with the space correlation function of the input, and integrating the result over the entire volume occupied by the element. If the input is plane wave "white" random noise, its correlation function will be an impulse. This greatly simplifies the integration, and by applying the Wiener-Khintchine theorem, results in the output power spectrum being proportional to the square of the absolute magnitude of the spatial transfer function. Spatial elements having random inputs can be considered to be incoherent.

6. SPACE-FREQUENCY EQUIVALENCE

When a Fourier relationship exists between the radiation pattern and the amplitude distribution across the aperture, the reciprocal relationship between aperture and pattern widths is displayed by similarity of the radiation patterns for isofrequency receivers of large spatial extent and wide-band receivers of small extent. The equivalence existing in a receiving array between its spatial configuration and the frequency configuration of the source is illustrated in Figure (H-5). As the spatial configuration is varied to improve the directionality, in the frequency case this corresponds to using a source having wide-band signals. If a continuous, uniform array is replaced by point sources spaced at one half wavelength intervals, then the continuous frequency distribution of the sources will be replaced by a set of discrete frequencies. When the element spacing (or frequency spacing) is made large, multiple major lobes result.

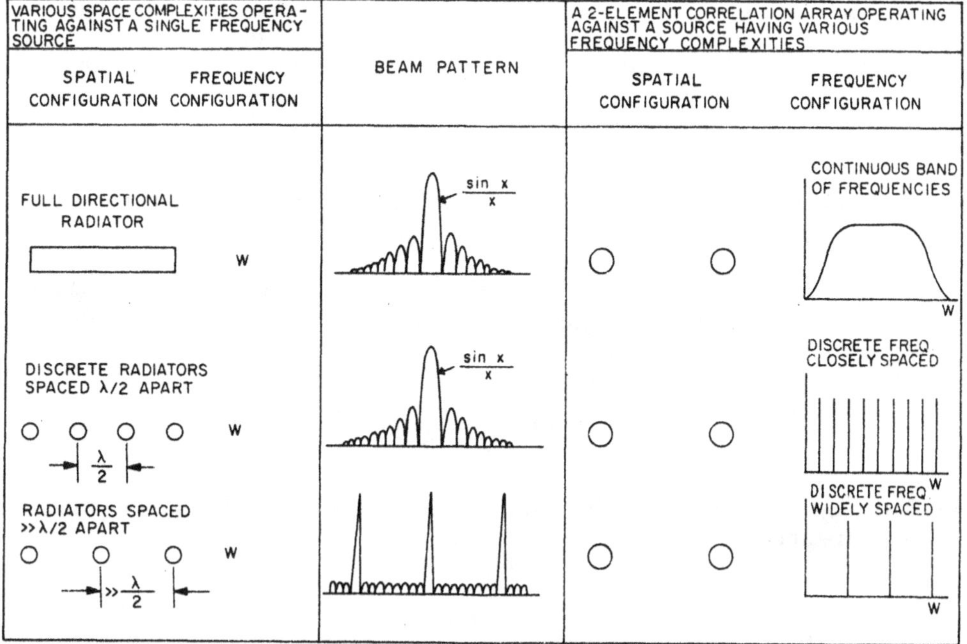

Figure H-5 - Equivalence between the complexity of the receiving array and the complexity of the frequency configuration of the source

7. LINEAR ELEMENT APPROXIMATION

INTRODUCTION

To approximate a spatial element means to approximate its point source response (or spatial transfer function) by a sum of predetermined responses, properly weighted and constrained. This, similar to circuit elements, may imply specifying the phase shift in a particular direction, or the plane wave response at broadside, or normalizing the integral square value of the point source response over a given spatial coordinate system.

To synthesize a spatial element which will have a specified directional characteristic, it is often convenient to deal in terms of line sources. The idealized concept of a true line source is useful for studying the directional characteristics of many physically realizable transducers. Equivalent line source concept facilitates evaluation of the directivity pattern in a single plane of more complicated transducers. For example, a cylindrical source with a radius less than 1/6 wavelength is closely approximated by a line source and its directivity pattern may be synthesized by means of line-source theory. Another case of practical interest is the plane-surface radiator where the directivity pattern in a single plane may be obtained by considering an equivalent line source.

Various methods for synthesizing line sources to obtain specified directivity patterns will be discussed. An examination will be made of the effects of finite aperture width and of amplitude and phase errors on the radiation pattern. In the discrete case, that is, a linear multi-element array, there is the additional consideration of departure from uniform spacing. These facets of the synthesis problem, including use of integral transforms and additional descriptions for evaluating the performance of spatial elements, are also included in the discussion.

TRUE RADIATION PATTERN

Before the true radiation pattern is considered in terms of the synthesis problem, additional insight will be obtained by reviewing the highlights of its derivation. In the transmission of acoustic waves, there is no rotational motion of the particles so that the velocity vector of a particle is an irrotational vector. If u, v, w are the velocities in the x, y, z directions, respectively, the velocity vector $\mathbf{V}$ may be represented as the gradient of some scalar potential function Ψ, i.e.,

$$\mathbf{V} = \nabla\Psi \tag{H-5}$$

where Ψ is termed the velocity potential. Equation (H-5) may also be written as

$$\left.\begin{aligned} u &= \frac{\partial\Psi(x,y,z,t)}{\partial x} \\ v &= \frac{\partial\Psi(x,y,z,t)}{\partial y} \\ w &= \frac{\partial\Psi(x,y,z,t)}{\partial z} \end{aligned}\right\} \tag{H-6}$$

The velocity potential, in one-dimension, at a point M due to a harmonic point source of strength Φ at a distance r is given by

$$\Psi(M) = \frac{\Phi}{4\pi r}\, e^{j(\omega t - kr)} \tag{H-7}$$

where

$\omega/k = c =$ velocity of propagation through medium,

$k = 2\pi/\lambda =$ the wave number which plays the same role in space coordinates as ω does in time coordinates,

$\lambda =$ wavelength which measures the length of one cycle in space just as the period T measures one cycle in time.

The source strength $\Phi(y)$ is the distribution of strength along the source, where it is assumed to be finite and continuous and has a finite number of finite discontinuities. Outside the interval of source dimensions, $\Phi(y)$ is assumed to be zero, and from Eqs. (H-6) and (H-7), has the dimensions of velocity-volume per unit length. It will be assumed that the time variation of the source-strength corresponds to the single angular frequency ω.

Consider a source of length (a) in the coordinate system of Figure (H-6). The sound pressure at the point M, $p(M)$, is given by $-\rho(\partial\Psi/\partial t)$ where ρ is the density of the medium, and the total pressure at point M is found by integrating along the line of the source:

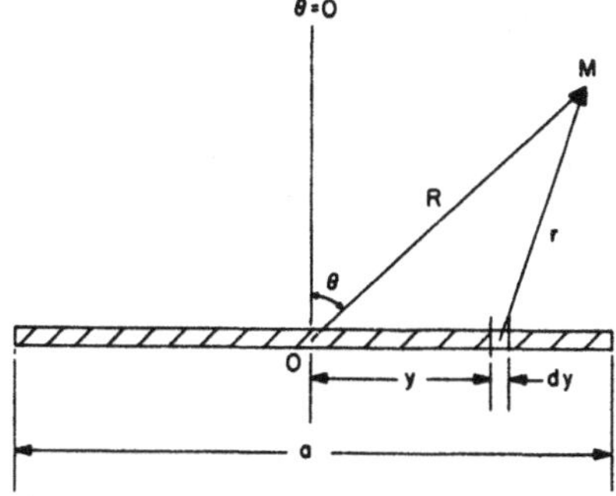

Figure H-6 - Coordinate system for directivity pattern of a line source

$$p(M) = -\frac{j\omega\rho}{4\pi} e^{j\omega t} \int_{-a/2}^{a/2} \frac{\Phi(y)\, e^{-jkr}}{r}\, dy. \qquad \text{(H-8)}$$

This equation holds for any point in space and is correct for any source-strength distribution for which the integral is convergent.

A requirement necessary for restricting point M to the Fraunhoffer diffraction region of the pressure distribution is that the path difference between contributions from the center and the end of the source be small compared to a wavelength. This may be expressed as

$$R \gg \frac{a^2}{8\lambda}. \qquad \text{(H-9)}$$

The acoustic pressures and particle displacements will then have common phases and amplitudes at all points on any plane perpendicular to the direction of wave propagation. Using Eq. (H-9), Eq. (H-8) becomes

$$p(M) = -\frac{j\omega\rho}{4\pi R} e^{j(\omega t - kR)} \int_{-a/2}^{a/2} \Phi(y)\, e^{jky \sin\theta}\, dy. \qquad \text{(H-10)}$$

Normalizing with respect to broadside $(\theta = 0)$, the normalized directivity pattern, $G(\theta)$, becomes

$$G(\theta) = \frac{p(\theta)}{p(0)} = \frac{\int_{-a/2}^{a/2} \Phi(y)\, e^{jky\sin\theta}\, dy}{\int_{-a/2}^{a/2} \Phi(y)\, dy}. \qquad \text{(H-11)}$$

If a normalized source strength distribution $F(y)$ is defined as

$$F(y) = \frac{\Phi(y)}{\int_{-a/2}^{a/2} \Phi(y)\, dy}, \qquad \text{(H-12)}$$

then Eq. (H-11) may be written as

$$G(\theta) = \int_{-a/2}^{a/2} F(y)\, e^{jky\sin\theta}\, dy. \qquad \text{(H-13)}$$

This is the standard equation for obtaining the directional response of a line source as a function of angle. Since $k = 2\pi/\lambda$, if $s = -\sin\theta/\lambda$, Eq. (H-13) becomes

$$G(s) = \int_{-a/2}^{a/2} F(y)\, e^{-j2\pi ys}\, dy. \qquad \text{(H-14)}$$

This is identical to Eq. (E-2) except for the finite limits of integration and states that in the far field, the normalized directivity pattern $G(\theta)$ is the finite Fourier transform of the relative source strength distribution $F(y)$. Sources with the same ratio of length to wavelength and with the same $F(y)$ will have the same directivity pattern. $F(y)$ is the normalized volume-velocity distribution per unit length across the aperture.

If we wish to have a pattern description independent of a/λ, and vary with θ only, define

$$z = \frac{a}{\lambda} \sin \theta . \qquad \text{(H-15)}$$

Then Eq. (H-13) becomes,

$$g(z) = \int_{-a/2}^{a/2} F(y)\, e^{j 2\pi \frac{y}{a} z}\, dy . \qquad \text{(H-16)}$$

Let $y/a = x/2$, then $dy = (a/2)dx$ and Eq. (H-16) may be expressed as

$$g(z) = \int_{-1}^{1} \frac{a}{2} F\left(\frac{ax}{2}\right) e^{j\pi z x}\, dx . \qquad \text{(H-17)}$$

If the weighted source strength distribution

$$\frac{a}{2} F\left(\frac{ax}{2}\right)$$

equals a new function of x, say $f(x)$, then we have,

$$g(z) = \int_{-1}^{1} f(x)\, e^{j\pi z x}\, dx , \qquad \text{(H-18)}$$

which is referred to as the true radiation pattern or "pattern function." The function $f(x)$ is termed the "excitation function" of the source. Note that (y/a) was chosen equal to $(x/2)$ and not (x) as might be expected, so that the limits in Eq. (H-18), the range of integration, is $(-1 \leq x \leq 1)$ and not $(-1/2 \leq x \leq 1/2)$. This corresponds to the angular interval $(-\pi/2 \leq \theta \leq \pi/2)$ which provides the "accessible" portion of the pattern function. It is so termed since it is the only portion of $g(z)$ that corresponds to a physically measurable value of the normalized sound pressure (the $g(z)$ is a single-valued nonperiodic function of z). Though $z > |a/\lambda|$ is possible, it corresponds to the "inaccessible" portion of the pattern and has significance for superdirective sources, that is, sources whose main beam is narrower than that from a source of the same length having uniform excitation.

Often, in trying to synthesize sources to obtain extremely narrow-beam directivity patterns, large minor lobes may occur in the inaccessible portion of the pattern. This is caused by large amplitude terms in the expansion of the excitation function $f(x)$ describing the source. Theory has shown that there is no upper limit to the gain of a radiator provided no limit is placed on the amplitude of the continuous excitation function or its derivatives.

An analysis in z-space is very convenient since two patterns generated by sources of different length but with the same excitation function are represented by the same function extending over different intervals. It should be remembered, however, that in the real physical domain, which may be characterized as θ-space, the two patterns appear as different functions over the same angular interval.

METHODS OF SYNTHESIS

The problem of synthesis is one of finding how to specify pattern functions with desirable properties in the accessible region and which can be achieved by practical excitation functions. There are two techniques which may be usefully applied to this problem and which result in approximate solutions. The first is a series expansion method whose accuracy depends on the number of terms considered. The second is based on the Fourier integral transform where the degree of approximation depends on the range of integration.

If we expand both $g(z)$ and $f(x)$ in a finite series of weighted elementary functions, i.e.,

$$g(z) = \sum_n a_n \psi_n(z) \tag{H-19a}$$

$$f(x) = \sum_n b_n \lambda_n(x) \tag{H-19b}$$

and require that the corresponding coefficients be equal for every value of the discrete variable $n(a_n = b_n)$, then substituting Eqs. (H-19a) and (H-19b) in Eq. (H-18) leads to the condition that

$$\psi_n(z) = \int_{-1}^{1} \lambda_n(x)\, e^{j\pi z x}\, dx . \tag{H-20}$$

The utility of functions satisfying Eq. (H-20) depends on whether the coefficients may be determined conveniently.

One possible technique is to form an orthonormal set of functions out of a known set of ψ_n's. If $g(z)$ is expanded as a finite sum of these functions, this will be the best approximation in the least squares sense. This is a desirable approximation if one is interested in maximizing the directivity factor of the source and can be the basis of synthesizing for maximizing the directional gain of a source. The ψ_n's themselves were not made orthogonal because it could not be assumed that any particular set of functions $[\psi_n(z)]$ will simultaneously satisfy the conditions of orthogonality and Eq. (H-20) for all values of a/λ.

Approximating in the least square sense is seen to be appropriate if it is desired to produce a specified signal on the principal axis with minimum energy in the sidelobes. This may be attributed to the finite Fourier series representation being the best approximation on the basis of energy. There are other criterion which may be used. For example, if one is interested in having minor lobes of low amplitude, approximation in the Tchebycheff sense may be more desirable. This permits distinguishing a weak source located on the principal axis from stronger sources located off the axis, and states that for a given number of sidelobes, the maximum value of a minor lobe may be made smallest if all lobes are of equal amplitude.

There are basically three sets of complementary functions which satisfy Eq. (H-20) and may be useful expansions of the pattern and excitation functions. One is where $f(x)$ is expanded in a Fourier series, while the other two provide power series or polynomial expansions of the excitation function.

If

$$\lambda_n(x) = \left(\frac{1}{2}\right) e^{jn\pi x},$$

then for a general type of line source, the pattern and excitation functions can be expanded in series of the form

$$g(z) = \sum_{n=-\infty}^{\infty} a_n \frac{\sin \pi(z+n)}{\pi(z+n)} \tag{H-21}$$

and

$$f(x) = \sum_{n=-\infty}^{\infty} \frac{a_n}{2} e^{jn\pi x} . \tag{H-22}$$

Equation (H-22) is recognized as the complex form of the Fourier series and infers that a continuous phase shift of the form $e^{jn\pi x}$ is applied to a uniformly excited line source. This, in effect, steers the main beam so that it is centered on the value $z = -n$ rather than on $z = 0$. The complete pattern is a superposition of $\sin z/z$ beams steered toward different directions in z-space. If n is allowed to exceed a/λ, the main portion of the nth beam will be steered into the inaccessible region of the pattern and only the "sidelobes" will contribute to $g(z)$ within the accessible region.

For an $f(x)$ that is finite and continuous over the full length of the source, its partial-sum expansion in terms of the λ_n's should converge in-the-mean. Then a realizable pattern function may be expanded in the corresponding set of ψ_n's.

Directional sources may also be synthesized by means of the Fourier integral transform utilizing the knowledge that in the far field, the excitation function and its corresponding pattern function are a Fourier transform pair, indicated without proof in section E. This method can then make use of the inversion properties of the Fourier transform, its limitation being that $g_1(z)$, the "desired" pattern function, can be specified in advance only over the accessible region of the pattern.

With reference to Eqs. (E-2) and (E-3), by letting $z = (a/\lambda) \sin \theta$, $(y/a) = (x/2)$, and $(a/2)F(ax/2) = f(x)$, for reasons given earlier, the equations become

$$g_1(z) = \int_{-\infty}^{\infty} f(x)\, e^{j\pi zx}\, dx\,, \tag{H-23}$$

$$f(x) = \int_{-\infty}^{\infty} g_1(z)\, e^{-j\pi zx}\, dz\,. \tag{H-24}$$

Equation (H-23) is equivalent to Eq. (H-18) except for the limits in integration. It will be remembered that if one member of a Fourier transform pair, say $g_1(z)$, is of finite length, the other, $f(x)$, will be infinitely long. If $g_1(z)$ is specified to be different from zero only in the accessible interval $(-a/\lambda \le z \le a/\lambda)$, $f(x)$ cannot be zero outside the interval $(-1 \le x \le 1)$ and the achieved pattern, $g(z)$, differs from the specified pattern, $g_1(z)$, by an "error pattern," $\Delta g(z)$, given by

$$\text{error pattern} = \Delta g(z) = g_1(z) - g(z)\,. \tag{H-25}$$

Due to the finite range of integration of Eq. (H-18), the Fourier transform method may be insufficient for pattern synthesis. However, it is of value in synthesizing sources with oddly shaped patterns and for problems where a series solution may prove too laborious.

EFFECT OF FINITE APERTURE WIDTH

The effect of limiting the size of the aperture is to introduce sidelobes in the radiation pattern. The more smoothly an aperture distribution goes to zero at the edges of the aperture, the smoother will be the radiation pattern. "Smoothness" in a radiation pattern implies not only an absence of sidelobes, but also an absence of sharply defined beams. This is illustrated in Figures (H-7) and (H-8). A finite aperture width implies sidelobes, which may be minimized but only at the expense of broadening the major lobe.

Two important aperture distributions (excitation functions) in spatial element analysis are the

Gaussian Distribution

$$f(x) = e^{-\pi x^2} \tag{H-26}$$

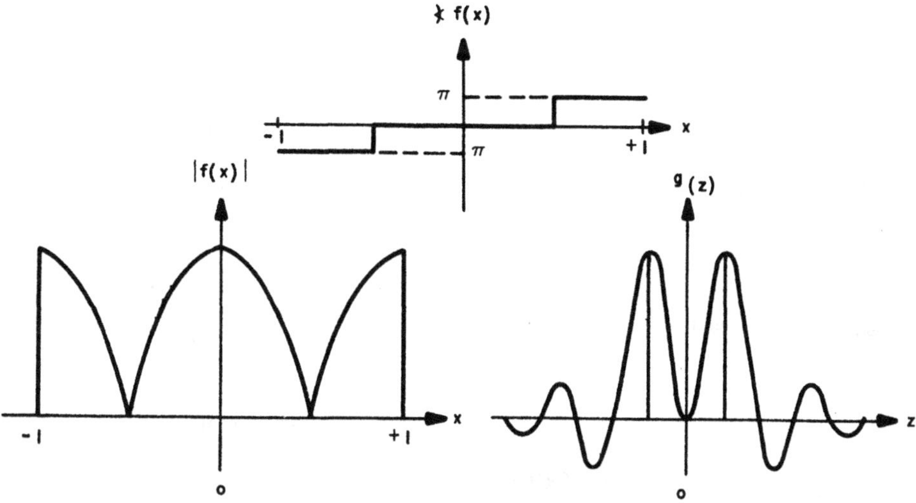

Figure H-7 - Examples of the relation between the field at the edges of the aperture and sidelobes in the radiation pattern

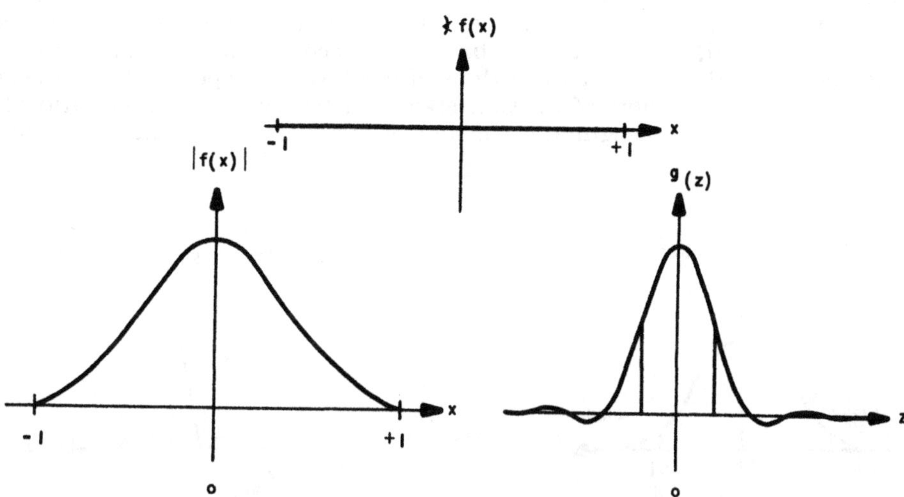

Figure H-8 - Examples of the relation between the field at the edges of the aperture and sidelobes in the radiation pattern

and

Rayleigh Distribution

$$f(x) = xe^{-\pi x^2} \tag{H-26}$$

These distributions possess the very useful property of having self-reciprocal Fourier transforms. That is, if the aperture is of infinite width, the radiation patterns will have the same form as the aperture distributions, with the variable x replaced by z (or s). The waveforms are shown in Figure (H-9). That the Rayleigh distribution is proportional to the derivative of the Gaussian distribution explains their relative properties.

In practical cases, the aperture width will not be infinite, the radiation pattern will develop sidelobes, and the reciprocity will be lost. Since a true Gaussian pattern has no sidelobes, a

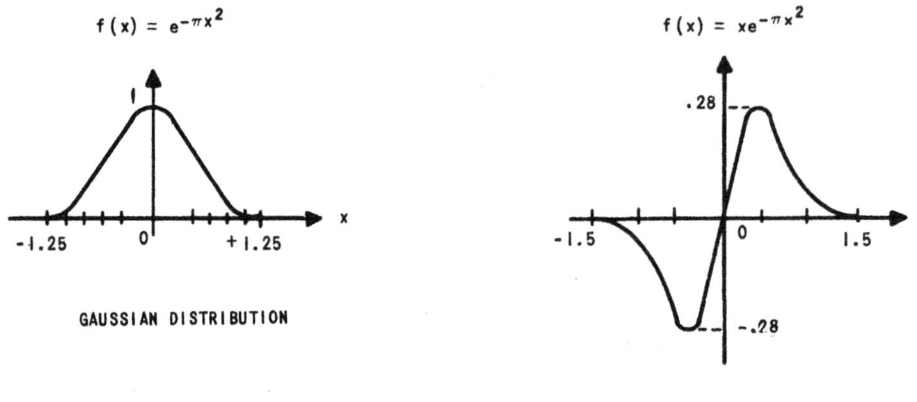

Figure H-9 - Self-reciprocal transforms

Gaussian-like distribution (distinguished by having most of its amplitude at the center of the aperture) is suited for producing a pattern with small sidelobes although this does not give the narrowest beamwidth for a given sidelobe level. In general, antisymmetrical patterns tend to approximate the Rayleigh distribution and can be interpreted in a similar way as was the Gaussian function. For a finite aperture, the Gaussian distribution and its corresponding pattern are given in Figure (H-10). Whether a broad or narrow Gaussian distribution is desired would determine applying either high or low values of excitation, respectively, to the edges of the aperture. A measure of the width of the Gaussian distribution is the taper ratio which is defined as the ratio of the field strength at the center of the aperture to that at its edge.

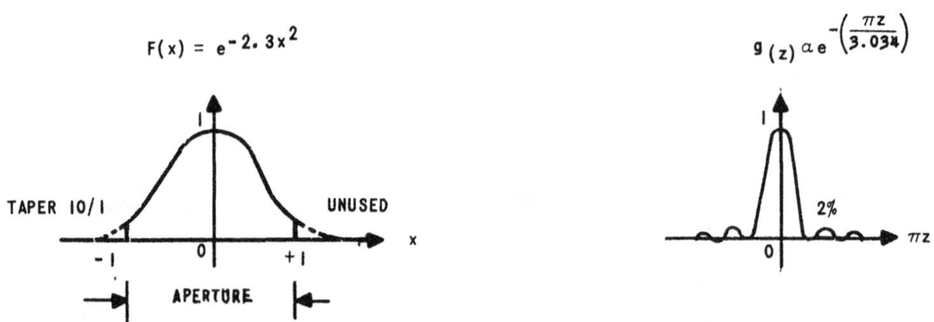

Figure H-10 - The effect of a finite aperture upon a Gaussian distribution and its corresponding radiation pattern

EFFECT OF ELEMENT SPACING

A method of transmitting sound or receiving sound unidirectionally is by means of a linear array of small nondirectional transducers referred to as point elements. If the elements are directional, the directive properties of the array may be determined by considering the directive properties of both the individual elements making up the array, and an array of isotropic radiators at the location of the point elements.

For the array of $(2N+1)$ arbitrarily spaced point elements shown in Figure (H-11), the discrete form of the pattern function, Eq. (H-11), is

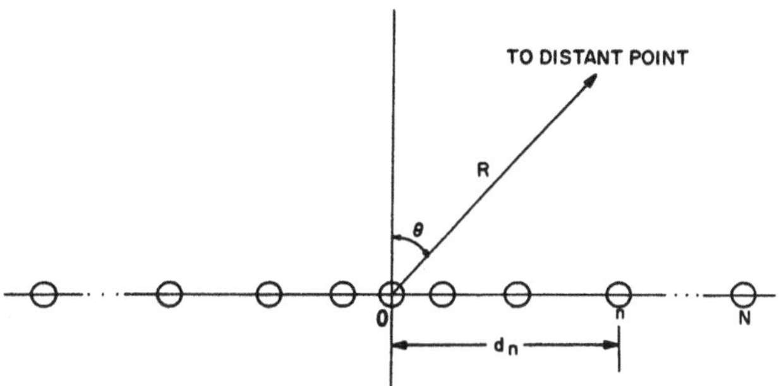

Figure H-11 - A linear array of 2N + 1 unequally spaced elements

$$G_{2N+1}(\theta) = \frac{p(\theta)}{p(0)} = \frac{\sum_{n=-N}^{N} \Phi_n e^{jkd_n \sin\theta}}{\sum_{n=-N}^{N} \Phi_n}, \tag{H-27}$$

where d_n is the distance from the nth element to the center of the array. If a relative source strength or excitation coefficient, b_n, is defined as

$$b_n = \frac{\Phi_n}{\sum_{n=-N}^{N} \Phi_n}, \tag{H-28}$$

then Eq. (H-27) may be written as

$$G_{2N+1}(\theta) = \sum_{n=-N}^{N} b_n e^{jkd_n \sin\theta} \tag{H-29}$$

Note that the directional characteristic is normalized for unity in the broadside direction, i.e.,

$$G_{2N+1}(0) = 1. \tag{H-30}$$

With symmetrical excitation,

$$b_n = b_{-n} \qquad (n = 1, 2, \ldots, N) \tag{H-31}$$

Eq. (H-29) becomes

$$G_{2N+1}(\theta) = b_o + 2 \sum_{n=1}^{N} b_n \cos(kd_n \sin\theta). \tag{H-32}$$

If the elements are uniformly spaced a distance d apart, then $d_n = nd$, and the sidelobe level and beamwidth may be controlled by varying the element excitation. Equation (H-29) may be regarded as a Fourier series expansion of the radiation pattern where the b_n's are determined in a least mean-square sense, i.e., are Fourier coefficients. A general result of uniformly spaced arrays is that the sidelobe level may be reduced by decreasing the aperture excitation toward the extremes of the array, as indicated earlier. This reduction is obtained at the expense of the array beamwidth.

If the elements are nonuniformly spaced, d_n will not be a rational multiple of some unit distance. Hence, the nonuniformly spaced array is characterized by spatial frequencies which are not related by integers and where the b_n's must be determined by a non-mean-square error criterion.

An unequally spaced array has many interesting properties. For example, there is an equivalence between amplitude tapering of a uniformly spaced array and the space variation in a nonuniformly spaced array. Hence, nonuniform element spacing may be used to reduce sidelobes. Use of perturbation methods can reduce the sidelobe level to about $2/N_T$ times the main lobe level, where $N_T = 2N + 1$ is the total number of elements, without increasing the beamwidth of the main lobe. To achieve this reduction implies retaining uniform excitation.

A perturbation analysis may be performed to indicate small nonlinear changes in element spacing. However, the more useful properties of nonuniformly spaced arrays depend on large nonlinearities in the element spacings. An approximate method for making an analysis of such arrays may be achieved by representing it with an equivalent uniformly spaced array (EUA). This is done by Fourier expanding each term in Eq. (H-29) into an infinite number of uniformly spaced equivalent elements and adding the individual expansions term by term according to the spatial frequency. For practical purposes, only a few terms of the expansion need be considered. The EUA is the best mean square representation for the original array. It does not physically exist but is used merely for analysis of unequally spaced arrays.

ADDITIONAL DESCRIPTIONS

In general, directivity, both in transmitting and on receiving, is dependent on the ratio of the sound wavelength to the dimensions of the radiator. If the wavelength is large compared to the dimensions, the sound is emitted uniformly in all directions and the transducer response will be independent of the direction of sound incidence. If the dimensions are large compared to a wavelength, the radiation energy, received or transmitted, will be directional. Useful measures of the directive properties of spatial elements are the directivity factor and directivity index.

The directivity factor (D.F.) is defined as the ratio of the intensity or mean square pressure of the radiated sound in a free field at a remote point on the maximum response axis (MRA) to the intensity or mean square pressure at the remote point averaged over all directions. The distance must be sufficiently great so that the sound appears to diverge spherically from the effective acoustic center of the source. The average intensity of the sound passing through a large sphere of radius r is found by integrating the normal component of the intensity I_n over the surface of the sphere and dividing by the area, $4\pi r^2$. The directivity factor may then be expressed as

$$\text{D.F.} = \frac{4\pi r^2 I_o}{\int_S I_n \, dS}, \tag{H-33}$$

where I_o is the intensity at the remote point on the MRA. In the far field, the intensity of the radiated sound is the square of the absolute value of the pressure divided by ρc. Equation (H-33) then becomes

$$\text{D.F.} = \frac{4\pi r^2 p_o}{\int_S p^2 \, dS}. \tag{H-34}$$

For a line source of length (a), symmetrically excited, the directivity factor may be written in the form

$$\text{D.F.} = \frac{a/\lambda}{\int_0^{a/\lambda} g^2(z) \, dz}. \tag{H-35}$$

The problem of increasing the directivity factor of a line source with a given ratio of a/λ is equivalent to decreasing the integral

$$\int_0^{a/\lambda} g^2(z)\, dz \tag{H-36}$$

subject to the normalization requirement that $g(0) = 1$. Using expansions (H-19a) and (H-19b) and condition (H-20), for a finite number of terms N, maximizing the directivity factor becomes one of minimizing the integral

$$\int_0^{a/\lambda} \left[\sum_0^N a_n\, \psi_n(z)\right]^2 dz \tag{H-37}$$

subject to the requirement that

$$\sum_0^N a_n\, \psi_n(z) = 1. \tag{H-38}$$

In general, the problem of minimizing Eq. (H-37) is done using the method of Lagrangian multipliers.

The directivity index (D.I.) is the expression of the directivity factor in decibels; thus,

$$\text{D.I.} = 10 \log_{10} \text{D.F.} \tag{H-39}$$

For a linear array of point elements, uniform excitation is necessary to produce the maximum directivity index (MDI) for element spacing of integral-half-wavelengths. The maximum directivity factor is then numerically equal to the number of elements N (even or odd) in the array, i.e.,

$$(\text{D.F.})_{max} = N \qquad d/\lambda = n/2 \qquad (n = 1, 2, \ldots). \tag{H-40}$$

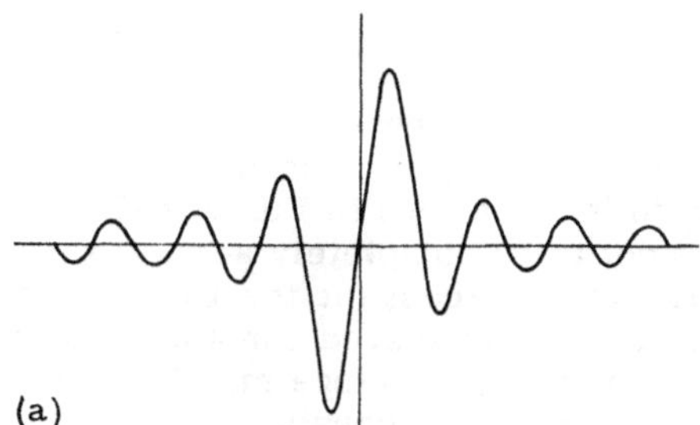

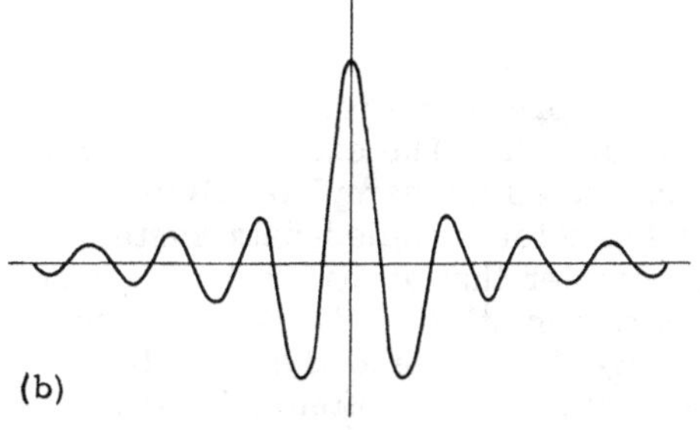

Figure H-12 - Rectangular plots of a typical (a) difference pattern and (b) sun pattern

For other values of element spacing the MDI is not obtained with uniform excitation. There is little difference between the MDI and the directivity index due to uniform excitation for element spacings greater than a half-wavelength. However, for $d/\lambda < 1/2$ there may be a significant improvement in the directivity index in going from uniform excitation to MDI excitation. The patterns due to the latter are superdirective and obtained only at the expense of requiring out-of-phase excitation and relatively large range of amplitudes.

Certain applications may require special types of directivity patterns, such as a difference pattern. Whereas the sum pattern exhibits even symmetry about a line drawn perpendicular to the radiator aperture at its midpoint the difference pattern exhibits odd symmetry about the same axis. A comparison is shown in Figure (H-12) for a line source. A sum pattern usually will have one major lobe in the direction of the principal axis while a difference pattern has two equal lobes with a null in the direction of the principal axis.

Combinations of sum and difference patterns are used in sonar and radar systems to improve the accuracy of bearing measurements. The error signal obtained is primarily determined by the slope of the difference pattern in the vicinity of the origin. To enhance the sensitivity to small changes in angle, it is necessary that the slope be as steep as possible. For sum patterns the influencing properties are directivity index, beamwidth and sidelobes. For difference patterns, slope and sidelobe level are significant for determining angular sensitivity. If the sidelobes are too large, false indications of target direction may result in the presence of multiple targets.

Consider the pattern function $g(z)$ in terms of the excitation function $f(x)$ for a line source,

$$g(z) = \int_{-1}^{1} f(x)\, e^{j\pi zx}\, dx\,. \qquad \text{(H-18)}$$

The slope of this pattern, with respect to z is given by

$$\frac{dg(z)}{dz} = \pi \int_{-1}^{1} jxf(x)\, e^{j\pi zx}\, dx\,. \qquad \text{(H-41)}$$

By setting $z = 0$, the slope at the origin is

$$\left.\frac{dg(z)}{dz}\right|_{z=0} = \pi \int_{-1}^{1} jxf(x)\, dx\,. \qquad \text{(H-42)}$$

If $f(x)$ is subjected to a constraint such as constant power radiated, that is,

$$\int_{-1}^{1} |f(x)|^2 dx = \text{a constant}, \qquad \text{(H-43)}$$

then by employing the method of the calculus of variations the excitation function which maximizes Eq. (H-42) may be determined. If the constant is arbitrarily set equal to one, a uniform-phase distribution function given by

$$f(x) = -j\ 1.22x \qquad \text{(H-44)}$$

will give rise to the pattern with maximum slope at the origin. Any uniform-phase distribution other than the linear distribution (H-44) will result in a smaller slope at the origin. The maximum slope pattern and the excitation function corresponding to it are shown in Figures (H-13) and (H-14), respectively. The term "uniform phase" is seen not to be completely accurate since the phase of the difference pattern changes by 180°. However, except for the 180° phase reversal, the phase of the pattern and corresponding distribution is considered constant. Since the linear excitation function gives rise to the maximum slope pattern, the slope may be used as a figure of merit with which to compare slopes of other uniform phase, constant power difference patterns. The maximum slope pattern may not be the most desirable pattern to use in that the sidelobe level is quite high. Thus, a compromise must be made between slope and sidelobe level for angular error sensitivity.

In section F, the Z-Transform was shown to be useful for expressing discrete signals just as the Fourier Transform was for expressing continuous signals. The excitation distribution in the discrete elements of a linear array may be considered as the sampled values of a continuous function. Known relations in Z-Transforms developed for sampled-data systems can be used to simplify linear array analysis. It was shown earlier that arrays may be represented mathematically by polynomials and that important characteristics of the radiation pattern, such as the location and level of sidelobes and the beamwidth, can be analyzed in terms of the properties of the polynomials. However, these are approximate and often quite tedious to determine since the polynomials cannot generally be put in closed form. By employing Z-Transform theory, the array polynomial can be expressed in closed form permitting characteristics to be determined more conveniently.

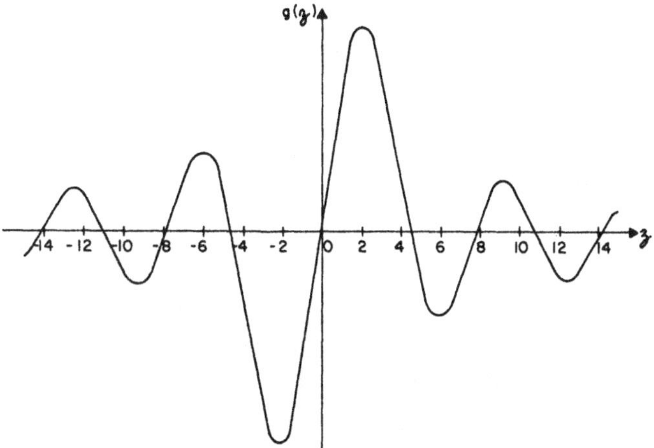

Figure H-13 - Maximum slope pattern

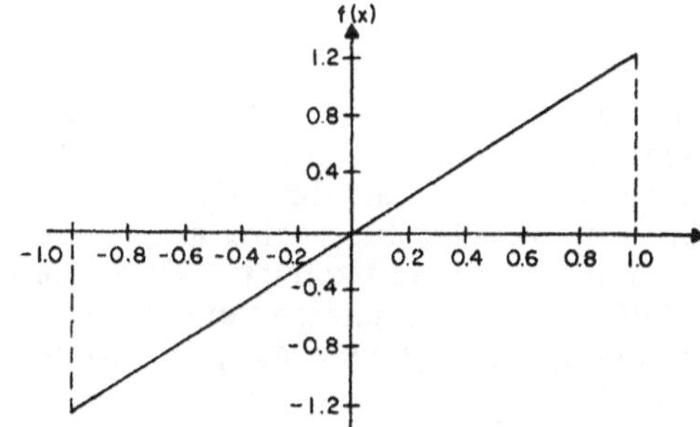

Figure H-14 - Excitation function corresponding to maximum slope pattern

Consider a linear array of N equally spaced elements. From Eq. (H-29), the polynomial for the pattern function associated with the array can be written as

$$G_N(z') = \sum_{n=0}^{N-1} b_n (z')^{-n}, \qquad \text{(H-45)}$$

where

$$z' = e^{-j\frac{2\pi d}{\lambda}\sin\theta} \qquad \text{(H-46)}$$

If the envelope of the amplitude distributions of the excitations in a linear array can be described by a continuous function $f(x)$ within the range $0 \le x \le (N-1)d$, then the excitation coefficients in Eq. (H-45) can be written as

$$\begin{aligned} b_0 &= f(0) \\ b_1 &= f(d) \\ &\vdots \\ b_{N-1} &= f\left[(N-1)d\right]. \end{aligned} \qquad \text{(H-47)}$$

Equation (H-45) then becomes

$$G_N(z') = \sum_{n=0}^{\infty} f(nd)(z')^{-n} - \sum_{n=N}^{\infty} f(nd)(z')^{-n}. \qquad \text{(H-48)}$$

For equal amplitude excitation in the two end elements,

$$G_N(z') = \left[1 \mp (z')^{-(N-1)}\right] F(z') \pm f(0)(z')^{-(N-1)}, \qquad \text{(H-49)}$$

where $F(z')$ is the Z-Transform of the function $f(x)$,

$$F(z') = Z\{f(x)\} = \sum_{n=0}^{\infty} f(nd)(z')^{-n}, \qquad \text{(H-50)}$$

$G_n(z')$ in Eq. (H-49) is expressible as a closed function of z', instead of a polynomial of N terms. Note that increasing the number of elements, N, in an array does not increase the complexity of the expression for $G_n(z')$.

EFFECT OF AMPLITUDE AND PHASE ERRORS

The design of an array requires that individual element amplitude and phase tolerances be maintained to achieve specified beamwidths, sidelobe levels, and difference pattern slopes. The effect of errors on the radiation function will be discussed briefly.

In general, the presence of errors in an aperture distribution (or excitation function) will cause some redistribution of directions in which the energy is radiated. This results in a reduction of the energy radiated along the main axis relative to the total radiation. If the errors vary slowly across the aperture, radiation components at small angles to the axis will develop, influencing the beamwidth and beamshape. Rapidly-varying errors will produce side radiation away from the main beam, but the increase in fine structure may not affect the radiation function appreciably.

Phasing errors generally affect the D.I., the sidelobe levels, and play an important role in determining bearing accuracy. If an array has the proper phasing in a specified direction, a random perturbation of the phasing will cause the intensity of the field and directivity index to decrease. Minimization of sidelobe levels may be attained by varying the vector amplitudes across the aperture in a predetermined manner, referred to as amplitude tapering. This is a slowly varying effect and results in changes in the beamwidth. If random phasing errors or amplitude errors are introduced in the excitation function, the symmetry necessary for minimization is destroyed and sidelobe levels increase.

REFERENCES

H-1. J. W. Horton, "Fundamentals of Sonar," United States Naval Institute, 1957, pages 170-176.

H-2. H. Unz, "Linear Arrays With Arbitrarily Distributed Elements," Electronic Research Laboratory, University of California, Series No. 60, Issue No. 168, Nov. 1956.

H-3. H. Unz, "Dimensional Lattice Arrays With Arbitrarily Distributed Elements," Electronics Research Laboratory, University of California, Series No. 60, Issue No. 172, 1956.

H-4. H. Gamo, "An Aspect of Information Theory in Optics," IRE Convention Record, Part 4, 1960, pages 189-203.

H-5. E. L. O'Neill, "Selected Topics In Optics and Communication Theory," Itek Corporation, Chapters II and III, 1958.

H-6. S. Matt and J. D. Kraus, "The Effect of the Source Distribution on Antenna Patterns," I.R.E. Proc., July 1955, pages 821-825.

H-7. P. M. Woodward, "A Method of Calculating The Field Over a Plane Aperture Required to Produce a Given Polar Diagram," J.I.E.E., Vol. 93, Part III A, 1956, pages 1554-1558.

H-8. H. S. Heaps, "General Theory for the Synthesis of Hydrophone Arrays," J. Acoust. Soc. Am., Vol. 32, No. 3, March 1960, pages 356-363.

H-9. W. E. Kock and J. L. Stone, "Space-Frequency Equivalence," I.R.E. Proc., Vol. 46, February 1958, pages 499-500.

H-10. S. S. Sandler, "Some Equivalence Between Equally and Unequally Spaced Arrays," I.R.E. Trans. on Antennas and Propagation, September 1960, pages 496-500.

H-11. R. F. Harrington, "Sidelobe Reduction by Nonuniform Element Spacing," I.R.E. Trans., Vol. AP-9, No. 2, March 1961, pages 187-192.

H-12. D. D. King, R. F. Packard, and R. K. Thomas, "Unequally Spaced, Broad-Band Antenna Arrays," I.R.E. Trans., Vol. AP-8, July 1960, pages 380-385.

H-13. D. K. Cheng and M. T. Ma, "A New Mathematical Approach for Linear Array Analysis," I.R.E. Trans. on Antennas and Propagation, May 1960, pages 255-259.

H-14. E. J. Powers, Jr., "Analysis and Synthesis of a General Class of Difference Patterns," Tech. Report No. 8, Research Laboratory of Electronics, Massachusetts Institute of Technology, July 30, 1959.

H-15. O. R. Price and R. F. Hyneman, "Distribution Functions for Monopulse Antenna Difference Patterns," I.R.E. Trans. on Antennas and Propagation, November 1960, pages 567-576.

H-16. J. L. Yen, "On the Synthesis of Line-Sources and Infinite Strip-Sources," I.R.E. Trans. on Antennas and Propagation, January 1957.

H-17. G. W. Swenson, Jr., "The University of Illinois Radio Telescope," I.R.E. Trans., Vol. AP-9 No. 1, January 1961, pages 9-16.

H-18. M. Leichter, "Beam Pointing Errors of Long Line Sources," I.R.E. Trans. on Antennas and Propagation, May 1960, pages 268-274.

H-19. L. E. Brennan, "Angular Accuracy of a Phased Array Radar," I.R.E. Trans. on Antennas and Propagation, May 1961, pages 268-275.

H-20. R. N. Bracewell, "Tolerance Theory of Large Antennas," I.R.E. Trans., Vol. AP-9, No. 1, January 1961, pages 49-58.

H-21. L. J. Chu, "Physical Limitations of Omnidirectional Antennas," J. Appl. Phys., Vol. 19, December 1948, pages 1163-1175.

H-22. S. A. Schelkunoff, "A Mathematical Theory of Linear Arrays," Bell Sys. Tech. J., Vol. 22, January 1943, pages 80-107.

H-23. J. B. Smyth, "Space Analysis of Radio Signals," J. Research N.B.S., Vol. 650, May-June 1961, pages 293-297.

H-24. W. H. von Aulock, "Properties of Phased Arrays," Bell Tel. System, Mono. 3779.

H-25. R. W. Bickmore, "A Note on the Effective Aperture of Electrically Scanned Arrays," I.R.E. Trans. on Antennas and Propagation, April 1954, pages 194-196.

H-26. R. N. Bracewell and J. A. Roberts, "Aerial Smoothing in Radio Astronomy," Aust. Jour. Phys., Vol. 7, December 1954.

H-27. T. T. Taylor, "Design of Line-Source Antennas for Narrow Beamwidth and Low Side Lobes," I.R.E. Trans., Vol. AP-3, January 1955, pages 16-28.

H-28. C. L. Dolph, "A Current Distribution for Broadside Arrays Which Optimizes the Relationship Between Beam-width and Sidelobe Level," Proc. I.R.E., Vol. 34, June 1946, pages 335-348.

H-29. G. Sinclair and F. V. Cairns, "Optimum Patterns for Arrays of Nonisotropic Sources," I.R.E. Trans., Vol. AP-1, February 1952, pages 50-61.

H-30. R. L. Mattingly, "Nonreciprocal Radar Antennas," I.R.E. Proc., Vol. 48, April 1960, page 795.

H-31. E. C. Jordan, "Acoustic Models of Radio Antennas," Ohio State University, Engineering Experiment Station Bulletin, No. 108, Columbus, Ohio, 1941.

H-32. L. L. Foldy and H. Primakoff, "A General Theory of Passive Linear Electroacoustic Transducers and Electroacoustic Reciprocity Theorem," J. Acoust. Soc. Am., Vol. 17, 1945, pages 109-120 and Vol. 19, 1947, pages 50-58.

H-33. T. G. Bell, "Hydrophone Minor Lobes Produced by Volume Scattering," J. Acoust. Soc. Am., Vol. 31, No. 10, October 1959, pages 1304-1307.

H-34. T. Morita, "Determination of Phase Centers and Amplitude Characteristics of Radiating Structures," Contract No. DA 04-200-ORD-273, Stanford Research Institute, Tech. Report 1-SRI Project 898, March 1955.

H-35. R. M. Wilmotte, "Correspondence: Note on Practical Limitations in the Directivity of Antennas," I.R.E. Proc., Vol. 36, No. 7, 1948.

H-36. R. Hills, Jr., "Synthesis of Directivity Patterns of Acoustic Line Sources," Tech. Memo. 23, Acoustics Research Laboratory, Harvard University, November 1, 1951.

H-37. R. L. Pritchard, "Directivity of Acoustic Linear Point Arrays," Tech. Memo. 21, Acoustics Research Laboratory, Harvard University, January 15, 1951.

H-38. "Calibration of Electroacoustic Transducers," American Standards Association, 1958.

H-39. D. R. Rhodes, "Introduction to Monopulse," McGraw-Hill, N. Y., 1959.

H-40. D. G. Tucker, "Space-Frequency Equivalence in Directional Arrays," Proc. of the I.E.E., Part C, March 1962.

I. CIRCUIT FILTERS

1. INTRODUCTION

Previous analyses have been concerned with descriptions of simple circuit and spatial elements — with a comparison of some of the analogous relationships which exist. The availability of numerous methods of describing these elements was noted along with the concept that the nature of the input played an important role in determining which description was to be employed. Additionally, the purpose of the analysis, that is, whether the description intended to describe a physical process, or facilitate computation in analysis or synthesis or to make the realization of a physical element in some sense easier or more economical, was also discussed. Practically, the elements described are subjected not to a single input — or even a single class of inputs — but to a wide variety. Some of the inputs contain information which should be preserved, and others discarded. In its broad sense, "filtering" represents an operation on the inputs in such manner as to discriminate against the interfering or undesired inputs while preserving the desired information. In view of the wide range of inputs — desired and undesired — and the range of functions, filters perform an impressive array of functions. Analysis procedures and instrumentation are constantly evolving. It is not proposed to review all of these in detail. Instead, discussion will be made of representative cases in order to illustrate in a sense, philosophical, rather than technical aspect of "filtering" operations. Although filtering represents perhaps the simplest operations with circuit elements, in determining the correct design, and in selection of the proper criterion for a particular application, it must be recognized that even for simple operations there may practically be complex compromises to reconcile.

A "classical" frequency filter is intended to separate two classes of signals whose spectra do not overlap. As has been previously discussed, time boundaries imposed on signals have spectra of large width and physically realizable filters cannot effect absolute separation. In practice, we try to make the ratio of output energies of the desired and undesired signals as large as possible. The classical filter specification does not take into account statistical properties.

There are two basic methods of designing such filters. The oldest method is based on image-parameter theory yielding Zobel filters. The other method is based upon insertion loss theory and gives the Darlington filters. Image-parameter methods are based on the study of elementary networks in terms of their image transfer constant and image impedances; the insertion loss method is based upon prescribed transmission characteristics. Though the insertion loss method is more involved, both theoretically and in computation, than the image method, it is not only more flexible but also a better approximation to the physical situation.

A class of filters of increasing importance are those necessary to separate a given signal from random noise whose spectrum overlaps that of the signal. In these cases, statistics of the signals plays an important role in the determination of the filter. It is necessary to select a suitable criterion and to determine how much noise may be accepted and how much signal energy may be rejected to achieve the desired result. This problem may be approached from two different points of view:

An <u>extraction filter</u> may be designed to recover or extract the message from a message-noise complex with minimum message distortion. A suitable criterion for this problem involves minimization of the rms difference between actual filter output and message. This problem was investigated by Wiener (1949) for a continuous, time-invariant, linear filter having infinite memory time (observation time) operating on a stationary random signal. Zadeh and Ragazzini (1950) then extended Wiener's work for a time-invariant linear filter, having specified memory time, for use with signals consisting of both a nonrandom polynomial and a stationary random component. They assumed the signal to be obscured by stationary random noise and their filter reduces to a Wiener filter in the case when the nonrandom part of the input is zero and the memory time is infinite. Work has also been done by Booton (1952) for time-varying filters, by Zadeh (1953) for nonlinear filters, and a host of others who have, in part, extended the theory to include discrete filtering. The solutions to these types of problems are integral equations that relate optimum filter characteristics to the statistics describing message and noise. Although the mathematics involved is complex, optimum filters can often be closely approximated by fairly simple apparatus, such as delay line filters.

A predetection filter may be designed to increase the possibility of detecting the presence of the message in the filtered output. A useful criterion for this problem is maximization of the signal (S) to noise (N) amplitude ratio:

$$\frac{S}{N} = \frac{\text{instantaneous peak signal amplitude}}{\text{rms noise amplitude}}$$

North (1943) investigated this problem for the case of additive white Gaussian noise and later, Zadeh and Ragazzini (1952) treated the more general problem of nonwhite noise. The determination of an optimum filter for nonwhite noise is usually quite complicated.

2. CLASSICAL FILTER

An image filter is a network made to operate out of and into appropriate impedances so that the conditions of maximum power transfer are approximated over the range of frequencies to be transmitted. Its transfer constant is a measure of the attenuation and change in phase encountered in transmission through the device. If the terminating impedances are selected in a consistent manner, then the overall image transfer constant of a cascaded group of image filters is the sum of their individual transfer constants. Thus, the characteristics of each element will contribute separately, and in a predictable manner, to the performance of the whole. This permits different functions to be designed as separate units which is very desirable in a complex system. However, this method has the disadvantage that the filter is assumed to be terminated in its image impedance while, in practice, the filter is generally terminated in a pure resistance. Since the image impedances vary widely with frequency, it is not possible to achieve an image termination at all frequencies in the passband. Consequently, reflections are set up at the terminals, and the attenuation, phase shift, and insertion loss will not be the same as computed on an image basis. A number of correction factors and additional matching networks are needed to achieve the desired transmission properties in actual operation with resistive terminations. Even then, the filter usually cannot be designed with a minimum number of circuit elements, and in some cases, the solution proves impossible.

The insertion loss method originated from early work of Bennett and Norton and was developed independently by Piloty, Darlington, and Cauer in 1939. The design of an insertion-loss filter is based upon prescribed transmission characteristics; it has been proven that for each of the effective transmission characteristics that is permissible for reactive networks, there exists a ladder configuration composed of simple elementary reactive 4-terminal networks which realizes it. This kind of design has become, in recent years, the principal method of filter design, in spite of the greater computational work it requires as compared to designing on an image basis. The modern network approach is more exact and leads to designs which are physically attainable in practice and whose final characteristics agree with calculations. There are a number of characteristics which are required. In the design of systems for pulse transmission, filters are often needed to fulfill requirements for both the attenuation behavior, and the phase or the group-delay behavior in the passband. To illustrate the flexibility of this method of design, several filters will be described:

(a) Filters Having Prescribed Attenuation Requirements

1. Power-term Filters— the attenuation behavior is represented, except for a constant, by a power series which corresponds to an attenuation curve beginning flat at $f = 0$ and rising monotonically to infinity.

2. Tchebycheff Filters — the attenuation characteristic is expressed, except for a constant, as a Tchebycheff polynomial, having a maximum slope in the transition between pass and stop bands. The attenuation versus frequency is allowed to oscillate or ripple between prescribed limits in the pass and reject bands while the phase and transient response are disregarded. Thus, this design is useful when only the amplitude characteristic is significant.

3. Butterworth (maximally-flat) Filter — this is actually a limiting case of the Tchebycheff design where the ripple in the passband is reduced to zero. Phase and transient response are considerably better than those attainable with the Tchebycheff design. The filter is characterized by considerable overshoot and undershoot when driven by a step function and is of primary value when a flat frequency response in the passband is desired.

(b) Filters Having Prescribed Phase Requirements

1. Maximally Linear Phase (Bessel) Filter — the time delay throughout the passband and most of the transition band is a constant. Thus, it has an excellent transient response with minimum overshoot; there is no region of constant amplitude in the passband. For a given number of filter elements, the slope in the transition region is much less than the Butterworth and Tchebycheff designs. This design is best suited for passing rectangular pulses or modulation envelopes and where overshoot or ringing is undesirable.

2. Transitional Butterworth-Thomson Filter — the characteristics are between those for the linear phase and maximally flat designs. Any degree of overshoot between the limits of the two designs can be selected as the controlled characteristic with the remaining characteristics being optimized. Rise time and transition slope will also lie between the limits of the two designs. This design is one of the best compromises between selectivity and transient response and usually results in excellent correlation between calculated and realized characteristics.

Although the insertion loss method is more involved, both theoretically and in computation, than the image method, it affords greater flexibility in physical problems. However, extensions of both methods have brought them closer together and towards a unified filter theory.

3. DETECTION OF A PERIODIC WAVE TRAIN

There are two basic methods for detecting repetitive signals upon which a strong ergodic noise signal has been superimposed. They are:

1. Correlation Analysis,
2. Comb Filtering

These methods are nearly equivalent, the selection of one over the other will depend upon the type of repetitive signal, the complexity of the instrumentation, and how the results are to be used.

CORRELATION ANALYSIS

The autocorrelation function $\psi(\tau)$ of the additive mixture of a repetitive signal $s(t)$ and a random noise $n(t)$ is

$$\psi_{S+N}(\tau) = \psi_{SS}(\tau) + \psi_{NN}(\tau) + \psi_{NS}(\tau) + \psi_{SN}(\tau) . \tag{I-1}$$

Assuming the mean values of both components to be zero, the crosscorrelation terms in Eq. (I-1) will vanish because of incoherence between signal and noise, and Eq. (I-1) simplifies to

$$\psi_{S+N}(\tau) = \psi_{SS}(\tau) + \psi_{NN}(\tau) . \tag{I-2}$$

Thus, the autocorrelation function of signal plus noise, both having zero means, is the linear superposition of the autocorrelation functions of each separately. In a region sufficiently remote from the origin, how far depending upon the frequency range of the random noise, the absence of a periodic signal is indicated by an autocorrelation function of constant (or zero) value whereas its presence will be evidenced by a periodic variation. The autocorrelation function for a sinusoidal signal in random noise is illustrated in Figure (I-1).

The improvement in signal-to-noise ratio in correlation equipment increases with the time of operation of the correlator. Therefore, theoretically an infinite signal-to-noise ratio can be obtained in the detection of a periodic signal in noise. However, in practical measurement, correlation must be determined in a finite time. A description of operation of a correlator may be given by statistical sampling theory. Figure (I-2) shows a portion of a random function whose autocorrelation is desired. Instead of shifting the function and performing a continuous multiplication, a set of samples $a_1, a_2, a_3, \ldots$ are taken, spaced at regular intervals as shown, and a second set of samples $b_1, b_2, b_3, \ldots$ is obtained, each sample trailing the corresponding sample in the first set by time τ_1. The autocorrelation curve at $\tau = \tau_1$, has the approximate value

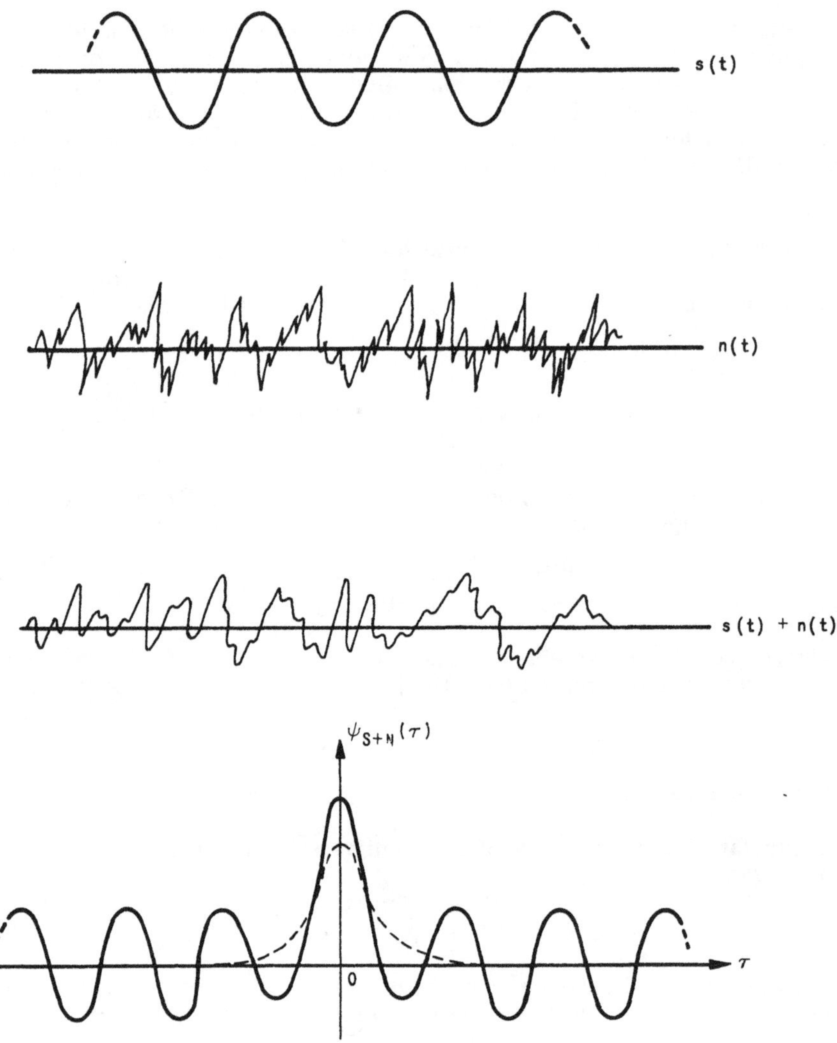

Figure I-1 - Use of autocorrelation to discover a signal in a strong background of noise

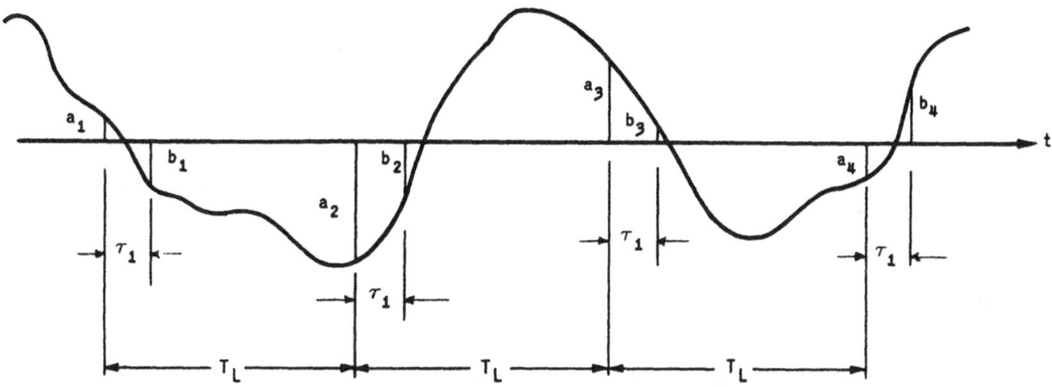

Figure I-2 - Determination of autocorrelation curve by statistical sampling theory

$$\psi(\tau_1) \cong \frac{1}{N} \sum_{n=1}^{N} a_n b_n . \tag{I-3}$$

If the sampling period T_L is sufficiently long, then the samples a_1, a_2, a_3, . . are practically independent of one another. The complete autocorrelation function is obtained by varying the spacing (τ) between samples. The accuracy in determining $\psi(\tau)$ and the improvement in signal-to-noise ratio increases with the number of samples values.

If the frequency of the repetitive signal is known, an even greater improvement in signal-to-noise ratio can be obtained by means of crosscorrelation. Using a reference signal of the same repetition period as the desired signal, the crosscorrelation function is

$$\psi_{S_1+N,S_2}(\tau) = \lim_{T \to \infty} \frac{1}{2T} \int_{-T}^{T} \left[s_1(t) + n(t) \right] \left[s_2(t+\tau) \right] dt . \tag{I-4}$$

$$= \psi_{12}(\tau) + \psi_{N2}(\tau) . \tag{I-5}$$

Since $n(t)$ and $s_2(t)$ are incoherent, $\psi_{N2}(\tau)$ vanishes, and unless the repetition frequencies of $s_1(t)$ and $s_2(t)$ are determined by the same source, $\psi_{12}(\tau)$ will also vanish. When $s_1(t)$ and $s_2(t)$ are incoherent some signal-to-noise improvement may be obtained by using a short-time approximation to $\psi_{12}(\tau)$, since the latter will not vanish. The improvement of signal-to-noise ratio is dependent on the time available which may be established by spectral fluctuations or by perturbations in the propagating medium.

When the reference signal $s_2(t)$ is comprised of a series of impulses whose period of repetition is coherent with that of the desired incoming signal, $s_2(t)$ may be expressed as the Fourier series

$$s_2(t) = \frac{K}{T} \sum_{n=-\infty}^{\infty} e^{jn\frac{2\pi}{T}t} \tag{I-6}$$

where K is the strength of the impulse and T is its period of repetition. Equation (I-5) then becomes

$$\psi_{12}(\tau) = \frac{K}{T} s_1(\tau) . \tag{I-7}$$

The result of crosscorrelation turns out to be precisely the desired signal $s_1(t)$ except for the magnitude factor K/T, which can be adjusted to any desired value by changing K. Therefore, crosscorrelation with a series of impulses having the same period of repetition as the signal,

1. removes the noise,

2. gives the desired signal without distortion,

3. gives the location of $s_1(t)$ without any unknown displacement of the origin.

The latter two results illustrate the advantage of crosscorrelation over autocorrelation since the latter generally distorts the signal and does not give its location in time.

COMB FILTERING

It was indicated in previous discussions that a periodic sequence of pulses could be represented by a line spectrum provided that the pulse train was not bounded in time — that is, existed over an infinite interval. When the pulse sequence was bounded, a continuous spectral distribution of energy resulted. A finite sequence of pulses may be represented by a continuous frequency spectrum consisting of a finite energy distribution concentrated at the frequencies where line components would exist if the sequence were infinite. The use of a "comb" filter which has passbands centered about harmonics of the pulse-repetition-frequency permits an improvement in output signal-to-noise in comparison to processing a single pulse. It should be recognized that signal-to-noise ratios alone do not serve as absolute performance criteria since such aspects as false alarms and incorrect dismissals are not directly indicated. However, signal-to-noise ratios when properly interpreted may be used at least in comparing some of the characteristics of various filtering operations.

Transfer functions of optimum comb filters may be determined by using the generalized methods of Zadeh and Raggazzini, to be discussed later. The physical operations associated with such filters consist of a cascade connection of a noise-shaping network, a single pulse filter, a nonfeedback type comb filter, and an output delay line. A representative configuration is shown by Figure (I-3). The transfer function of the noise-shaping network is equal to the reciprocal of the noise power spectrum $P_n(\omega)$. The purpose of this network is to preferentially weight the components of the signal where the noise spectrum has its lowest values. Following this weighting network is a single pulse filter which is identical with a North filter which represents optimum processing of a single rectangular pulse masked by white noise. The comb filter sums the individual pulses of the pulse train and weights them in proportion to their amplitudes.

The passbands of the filters are centered about multiples of the pulse-repetition-frequency, and the spread of the signal energy is inversely proportional to the number of pulses in the

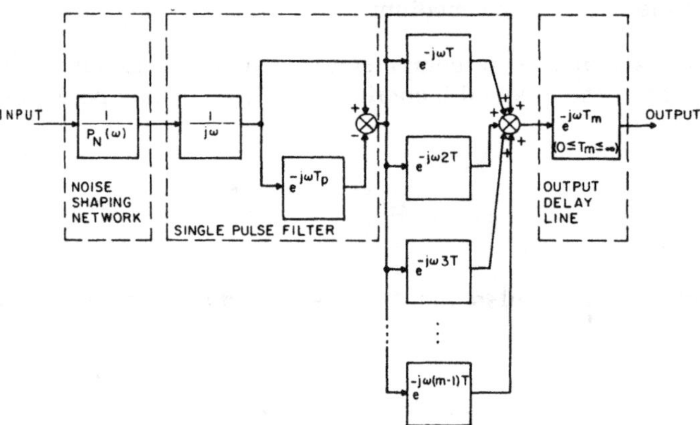

Figure I-3 - Filter maximizing signal-to-noise ratio at or after the trailing edge of the last pulse of a uniform pulse train

pulse train. When the width of the passbands is $2/mT$, with m equal to the number of pulses and T being the period, then approximately 90 percent of the signal energy about each prf is within the passband of the filter. For a small number of pulses, the distribution is broad and there may be an overlap of energy among the elements of the comb. However, the error in neglecting this overlap is small when the number of pulses is greater than 10. An approximate analysis of improvement to be expected by the use of comb filters involves considering the effect of band-limited noise which is added to the signal pulse train. It is assumed that the noise is gated and thus exists only for the duration of the pulse train. Consequently, noise energy rather than noise power can be used, having total mean energy N_o in the interval $-(2\pi\delta) \leq \omega \leq (2\pi\delta)$. The input signal-to-noise energy ratio is denoted by

$$r_i = 10 \log \frac{E}{N_o} \text{ (db)} \tag{I-8}$$

where E = total input signal energy. The filtered signal energy is approximately $0.90aE$ with (a) representing the filter gain. The noise output from the filters is

$$N_{cf} = a\left(\frac{2}{m} + \frac{\delta}{mf}\right) N_o . \tag{I-9}$$

Practically, $(\delta/f) << 2$ and N_{cf} may be approximately represented by $a2N_o/m$. Finally, the energy signal-to-noise-ratio at the output is given by:

$$r_o = 10 \log \frac{0.9E}{2\frac{N_o}{m}} = r_i + 10 \log 0.45m . \tag{I-10}$$

Consequently, the improvement is related to the number of pulses, with, of course, the requirement that the number of pulses to be processed be known and that the filter bandwidths conform to the pulse train. Figure (I-4) shows the improvement in output signal-to-noise-ratio where the comb elements are weighted uniformly. Other weights may be applied to the different elements of the combs. If the weighting is determined by applying the North matched-filter technique over the entire interval then the improvement is still dependent on the number of pulses plus a constant improvement of 4.5 db. An intermediate method has the filter gains adjusted to the envelope, that, the gain of the filter straddling the nth prf region would be proportional to the maximum amplitude of the signal spectrum at $\omega = 2n\pi/\tau$. This method involves an output signal-to-noise relationship given by $r_i + 10 \log 0.45m + 2.1$ db.

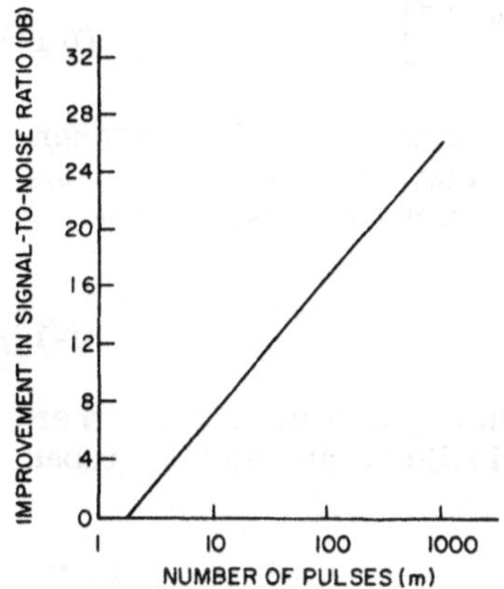

Figure I-4 - Improvement in signal-to-noise ratio due to a uniform comb filter as a function of the number of pulses in a sequence

The effect of the comb filter on the pulse shape will depend on the weights which are applied. The first method (corresponding to the uniform weighting of elements in an array), will have the least effect — with the pulse being rounded off somewhat. The North filter having the greatest improvement, will also have the greatest effect on the shape, producing an output which is almost triangular. Although the improvement in output signal-to-noise ratio should also improve the precision of localization in instances where the exact location is dependent on the pulse shape, for example on the leading edge of the pulse, comb filtering may introduce significant errors.

Detailed analyses of a number of realizable methods of comb filtering have been made in Refs. I-31 and I-32. In particular, a comparison is made of correlation with filtering methods. This analysis, using peak signal-to-rms-noise as the criterion, shows that cross correlation is equivalent to the optimum filtering briefly described in the preceding paragraphs. The correlator performs in the time domain an operation similar to the operation of the comb filter in the frequency domain. The comb filter passes harmonically related frequency components, while suppressing bands of frequencies lying between the passbands. Correlation involves multiplication of the incoming signal with a reference signal and averaging of the multiplier output with a low-pass filter. Only those frequencies present in both the input and the reference signal result in a zero frequency multiplier output component that passes through the low-pass filter. Decreasing the cutoff frequency of the low pass correlator filter corresponds to narrowing the width of the comb filter passbands. If crosscorrelation is employed, the reference signal is locally generated without noise. For autocorrelation, the reference signal is the input signal delayed by one pulse-repetition period and consequently is perturbed by noise — and as a result, autocorrelation is inferior to crosscorrelation. If the starting point of the incoming signal is not accurately known, then several correlation channels each having a different value must be employed, or it is necessary to store the signal and search through a range of (τ) values with the single channel. However, the adverse effects on comb filters caused by impulsive disturbances may be greater than for the correlator.

Although more detailed analyses of processing methods may be made, the interpretations of the analyses must be made with care. Output signal-to-noise ratios which have been used in establishing performance characteristics may inadequately describe the methods when employed in a system. Additional considerations which may be more difficult to handle analytically involve determining false alarms, incorrect dismissals, particularly when the interference may consist of disturbances having non-Gaussian characteristics.

COMPARISON OF ANALOG AND BINARY INTEGRATION

The detection of repetitive signals in noise may be improved by integration techniques. Regardless of the method used in any particular integration scheme, a fundamental requirement which they all have in common is that of a suitable memory. This memory must be able to accept and remember with sufficient accuracy a number of signals contaminated by noise. When a number of such sequences have been added while stored, their sum may then be extracted and examined.

The method for obtaining the desired signal storage may be either analog or digital in nature. Many of the basic analog integrating devices integrate by remembering the waveform of the signal and by using successive samples to obtain improvement. For example, if a succession of impulses are applied at intervals Δt to a single RC network, then the response E_{out} at the time the nth signal is applied is

$$E_{out} = \frac{1}{RC}\left[E_n + E_{n-1}e^{-\Delta t/RC} + \cdots + E_1 e^{-(n-1)\Delta t/RC}\right]. \tag{I-11}$$

The law of addition here is a weighted linear one in which the effect of each signal is exponentially weighted. When periodic signals are applied to regenerative delay-line integrators, the delay is made equal to the repetition interval of the signal and the summation again follows the law

$$E_{out} = E_n + E_{n-1}e^{-a} + \cdots + E_1 e^{-(n-1)a} \tag{I-12}$$

where a is the attenuation in nepers and is greater than zero. Extending this method to other elements, RC networks, regenerative dealy-line loops, narrow-band filters, and storage tubes, all have the same general law of addition, i.e.,

$$E_{out} = \sum_{q=1}^{n} E_q e^{-(n-q)\gamma} \tag{I-13}$$

If we were to calculate the output probability distribution when n mixed signals are added in accordance with Eq. (I-13), we would find that the signal-to-noise improvement is a function of n, the number of signal-plus-noise additions, and γ, the exponential weighting factor. Specifically, when successive repetition intervals are added, the noise increases roughly as the square root of the number of additions, and the signal increases as the number of additions. Hence, the relative signal-to-noise ratio should increase as the square root of the number of samples added. There are several disadvantages to analog integrators. The most important is that they require a large number of memory elements to store the waveform of the signal, thus making it difficult to realize long memory times.

Binary integration requires fewer memory elements since signals are quantized into two amplitude levels and in time between fixed time markers. In the process of quantizing, if the complex signal and noise waveform between given time markers exceeds a predetermined amplitude, a standard pulse is generated at the end of the interval. If the threshold is not exceeded, no pulse is generated. The probability of obtaining a standard pulse can then be determined from the probability distribution function for the given complex waveform. This method of integration then becomes a process of adding signal waveforms in successive repetition intervals.

4. PREDETECTION FILTER

INTRODUCTION

The primary purpose of a predetection filter is to enhance the strength of the signal relative to that of the noise and thereby facilitate detection. The form the filter takes will depend upon the information about the signal and noise that is available.

In most practical situations, information available is incomplete and it is necessary to make assumptions regarding the character of the noise and to select an adequate criteria from the standpoint of accuracy and convenience. Two types of predetection filters are the North or "matched" filter and the Zadeh-Ragazzini optimum predetection filter. Though the matched filter is actually a limiting case of the latter, it will be discussed separately due to North's theory pioneering the field of optimum filters. It also provides a good foundation for evaluating more complex predetection filters.

MATCHED FILTER

The correlation of one waveform with another can be carried out by passing the first waveform through a linear system whose impulse response is the time reverse of the second waveform and observing the output at a certain instant of time. If the two waveforms are identical, the filter is said to be "matched" to the input waveform. The filter output as a function of time is then the autocorrelation function of the waveform. Generally speaking, to distinguish among a group of signals (including the absence of a signal) masked by additive white Gaussian noise is equivalent to a coherent detection in which integrals I_k of the form

$$I_k = \int_{-\infty}^{\infty} y(t)\, x_k(t)\, dt \tag{I-14}$$

are compared with each other for given thresholds. In these integrals, $y(t)$ is the received signal and the $x_k(t)$ are the various signal waveforms in the absence of noise. If the integral is computed by multiplication and integration, the detection process is called "correlation detection." If the integral is obtained as the output of a linear filter at a given time, the process is then referred to as "matched filter detection." The two processes are, in a sense, equivalent.

To know whether a signal plus noise or just noise alone is present at a certain instant of time, say $t = t_o$, we require the filter output at that time to be greater when $x(t)$ is present than if it were absent. This is usually accomplished by making the instantaneous power in the filter output containing a signal at $t = t_o$ as large as possible compared to the average power in the noise at that time. If a mean-square criterion is used, for the case of additive white noise the signal-to-noise ratio ρ in the filter output may be expressed as

$$\rho \leq \frac{E}{N_o} \qquad \text{(I-15)}$$

where E is the total energy in the signal and N_o is the power spectrum of the noise, and is a constant. The equality in Eq. (I-15) is obtained for a filter whose impulsive response has the form of the image of the signal to be detected. That is, if $H(j\omega)$ is the complex transfer function of the element, then ρ is a maximum at time t_o for a signal $x(t)$ when

$$H(j\omega) = X^*(j\omega)\, e^{-j\omega t_o}. \qquad \text{(I-16)}$$

The transfer function is the complex conjugate of the Fourier spectrum of the signal multiplied by a phase factor $\exp(-j\omega t_o)$. Equation (I-16) is referred to as the Fourier transform criterion and the filter is called a matched filter.

Although the transfer function depends upon the instant of observation t_o, the corresponding value of the maximum ratio is independent of time and will thus be the same for all values of time for which $H(j\omega)$ satisfies Eq. (I-16). For ρ_{max} to be valid at any time t_o desired, we must obtain a physically realizable filter when t_o is inserted in Eq. (I-16). The necessary condition for realizability when dealing with real signals is that all of $x(t)$ must have entered the filter before the time t_o when the filter is expected to give maximum signal-to-noise ratio.

When Eq. (I-16) is satisfied, the output signal $y(t)$ will be, using Eq. (G-21),

$$y(t) = \frac{1}{2\pi} \int_{-\infty}^{\infty} |X(j\omega)|^2 \, e^{j\omega(t-t_o)} \, d\omega. \qquad \text{(I-17)}$$

This however, from the Wiener-Khintchine theorem, is the finite autocorrelation function of the input signal displaced by the time t_o. Therefore, the results of correlation analysis on arbitrary signal waveforms mixed with white, Gaussian noise may be deduced from the theory of matched filters.

In a matched filter, the product of the "widths" of the matched-filter output waveform and associated spectra should be a constant of the order of unity, the exact value of which depends on the definition of "width" (see section A.II-3). This means that the width of the signal component at the matched filter output cannot be less than the order of the reciprocal of the signal bandwidth. In simple detection problems, for the case of white, Gaussian noise, all signals having the same energy content are equally effective. Peak power, time duration, bandwidth and waveshape of the signal, per se, do not affect the output signal-to-noise ratio. However, for the case of bandlimited white noise of fixed total power, of all signals with the same energy, the one with the largest bandwidth is the most desirable. In system applications such as sonar and radar, it is often necessary to include requirements for range accuracy, resolution, and ambiguity, in addition to detection under noise-limited conditions. The idealized requirements of accuracy and resolution dictate a large bandwidth, while minimizing ambiguity requires a peak in the output of the matched filter at the time corresponding to the unknown delay and zero everywhere else. The width of the peak must be sufficiently small for multitarget and multipath situations.

If in addition to being delayed, the signal is also shifted in frequency by doppler effects, the receiver should contain a bank of matched filters. The detectability of the signal is still governed by the signal-to-noise ratio, Eq. (I-15), obtained without doppler shift. For multiple targets, each target represented at the receiver input should excite only the filter in the matched filter bank which corresponds to the target doppler shift (velocity) and should cause a sharp peak to appear in this filter's output envelope only at a time corresponding to the delay of the target, and nowhere else. The response of a filter at time (t) to a nondoppler-shifted, nondelayed signal, when matched to the doppler-shifted signal, is the real part or envelope of the complex Fourier transform

$$x(t,\phi) = 2 \int_{-\infty}^{\infty} F(j2\pi f)\, F^*\left[j2\pi(f-\phi)\right] e^{j2\pi f t}\, df \tag{I-18}$$

for $f > 0$, where ϕ is the doppler shift and $F(j2\pi f)$ is the Fourier spectrum of the received signal. Equation (I-18) is the joint autocorrelation function or ambiguity function, and for signal detectability its envelope $|x(t,\phi)|$ is required to be large at $t = 0$ if $\phi = 0$, and small otherwise. By applying the time-bandwidth product relationship to the ambiguity function, we find that the "width" of the peak response cannot be less than the order of $1/TW$. Thus, in order to obtain a very sharp central peak it is necessary, but not sufficient, to make the TW product of the signal very large.

In practice, single target conditions are not encountered, making it necessary to consider the relationships among signals. Specifically, the various signals should be distinguishable so that the overall probability of error in reception is minimized. For the special case of binary transmission, it turns out that if the two signals are a priori equally probable, one should use equal-energy anti-podal signals, i.e., $x_1(t) = -x_2(t)$. For the band-pass case in which the carrier phases are unknown, an optimum system is one where the signals are "envelope-orthogonal." Another method is using signals which are rectangular bursts of sine waves, the sine-wave frequencies of the different signals being spaced apart by integral multiples of $1/T$ cps, where T is the duration of the bursts. A third method is using code symbols in the form of orthogonal wide-band signals having the same energy per symbol. It is not essential that the symbols be strictly orthogonal but only that the interaction energy be small and not concentrated.

Since approximately $2WT$ "numbers" are sufficient to describe a signal which has an effective time duration T and an effective bandwidth W, a filter can be synthesized by $2WT$ elements or parameters. A form of matched filter is the tapped-delay-line filter. First, to illustrate the properties of matched filters consider the system of Figure (I-5). A signal, $x(t)$, for some duration T, may be considered to be generated by applying a unit impulse at $t = 0$ to a linear filter whose impulse response is $x(\tau)$. To this is added white noise $n(t)$ of power density N_o. The total signal $y(t) = x(t) + n(t)$ is then passed into a filter, matched to $x(t)$, whose output is denoted by $g(t)$. For the class of signals where the signal-generating filter can be represented as the tapped-delay-line spectrum shaper of Figure (I-6), the spectrum $X(j2\pi f)$ of the signal will have the form

$$X(j2\pi f) = F(j2\pi f) \sum_{i=0}^{n} G_i(j2\pi f)\, e^{-j2\pi f \Delta_i} \tag{I-19}$$

where Δ_i is the delay associated with the (i)th terminal. A filter matched to this signal may be obtained by replacing the G_i's and $F(j2\pi f)$ by their complex conjugates and applying the input at the end of the delay time, i.e., at the terminal Δ_n. The transfer function $H(j2\pi f)$ of the tapped-delay-line matched filter will then be

$$H(j2\pi f) = F^*(j2\pi f) \sum_{i=0}^{n} G_i^*(j2\pi f)\, e^{-j2\pi f(\Delta_n - \Delta_i)} \tag{I-20}$$

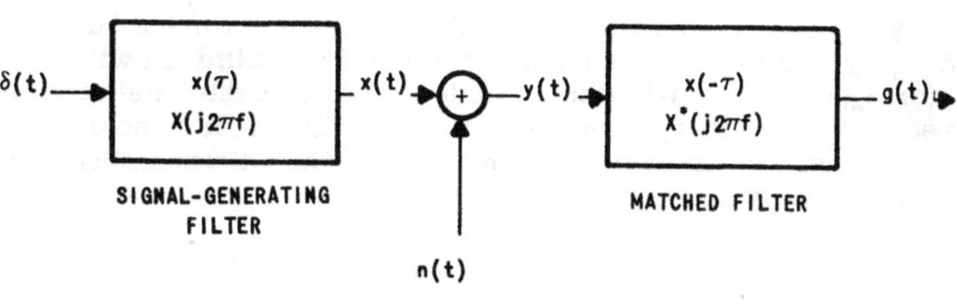

Figure I-5 - Illustrating the properties of matched filters

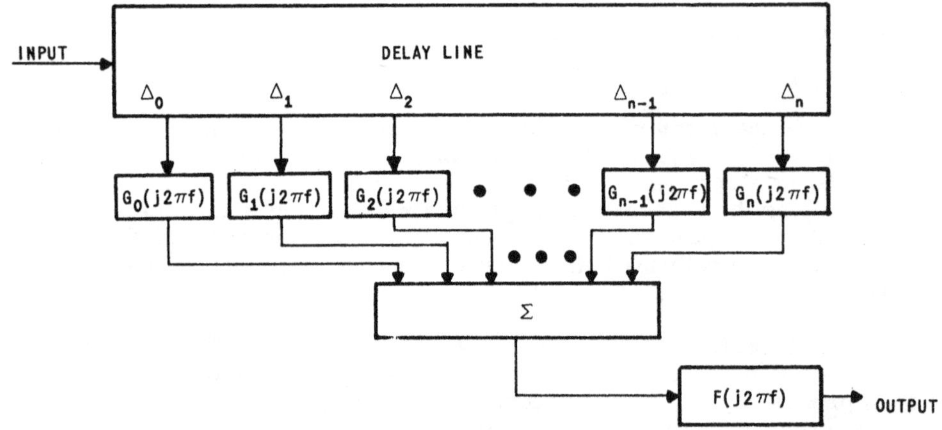

Figure I-6 - A tapped-delay-line spectrum shaper

which, when compared with Eq. (I-19), becomes

$$H(j2\pi f) = X^*(2\pi f)\, e^{-j2\pi f \Delta_n} \tag{I-21}$$

If $F(j2\pi f)$ and $G_i(j2\pi f)$ are assigned phase functions which are uniformly zero, then $F^*(j2\pi f) = F(j2\pi f)$ and $G_i^*(j2\pi f) = G_i(j2\pi f)$. The advantages of a single filter which can perform signal generation and matching is quite evident in situations where the transmitter and receiver are physically at the same location. The various restrictions set by a problem may be accounted for by adjusting the characteristics, i.e., $F(j2\pi f)$, $G_i(j2\pi f)$ and Δ_i for all (i), of the filter.

ZADEH-RAGAZZINI OPTIMUM PREDETECTION FILTER

In evaluating the performances of various elements, it is necessary to note both the differences and similarities between the criteria used. First, let us consider a conventional filter (a network whose function is to separate signal from noise) whose input is the sum of a signal $s_i(t)$ and a noise $n_i(t)$. If the filter is linear so that the output is the sum of its responses to $s_i(t)$ and $n_i(t)$, denoted by $s_o(t)$ and $n_o(t)$, respectively, and if the output $s_o(t) + n_o(t)$ is required to be as close as possible to the input signal $s_i(t)$, then the filter is said to be optimum when using the mean-square error criterion if

$$\left\{\frac{1}{T}\int_0^T \left[s_o(t) + n_o(t) - s_i(t)\right]^2 dt\right\}^{1/2} = \text{a minimum}. \tag{I-22}$$

The only assumption made is that $s_i(t)$ and $n_i(t)$ are stationary and independent.

A predetection filter intends to facilitate the detection of $s_i(t)$, rather than to reproduce $s_i(t)$. Consequently, a predetection filter is said to be optimum if the "distance" between the signal component $s_o(t)$ and the noise component $n_o(t)$ is maximized with respect to a constraint on $n_o(t)$ (or $s_o(t)$). A constraint is needed, otherwise the "distance" could be made as large as desired merely by increasing the gain of the filter. If a mean-square error criterion is used, and the constraint is expressed in terms of the "distance" between $n_o(t)$ and the zero signal, the quantity to be maximized by the filter is

$$\left\{\frac{1}{T}\int_0^T \left[s_o(t) - n_o(t)\right]^2 dt\right\}^{1/2} - K\left\{\frac{1}{T}\int_0^T n_o(t)^2\, dt\right\}^{1/2} = \text{a maximum} \tag{I-23}$$

where K is a constant (Lagrangian multiplier). If the signal $s_i(t)$ is assumed to be a specified but otherwise arbitrary function of time, and the noise $n_i(t)$ is ergodic and has a known correlation function $\psi_n(\tau)$, then the time averages in Eq. (I-23) may be replaced by ensemble averages, with (t) held constant at a fixed value t_o, relative to a temporal frame of reference attached to the signal $s_i(t)$. Equation (I-23) then becomes

$$s_o^2(t_o) - \overline{\mu n_o^2(t)} = \text{a maximum} \tag{I-24}$$

where (μ) is a constant equal to $(K-1)$ and the bar indicates a time average (which equals the ensemble average). For a linear filter, Eq. (I-24) is equivalent to maximizing the signal-to-noise ratio (ρ), i.e.,

$$\rho = \frac{s_o^2(t_o)}{\overline{n_o^2(t)}} = \text{a maximum} \tag{I-25}$$

which is the criterion used in North's theory. Thus, if the criterion used to optimize a predetection filter is of the mean-square-type, and the filter is linear, then the filtering criterion reduces to the North criterion. The primary difference is that the noise in the North filter was assumed to be white, Gaussian noise, while in the present case, the only restriction imposed is that the noise be ergodic. The predetection filtering criterion, Eq. (I-23), may also be expressed as

$$\overline{n_o^2(t)} - K s_o(t_o) = \text{a minimum} \tag{I-26}$$

which is the most convenient form for design purposes.

For the case of nonwhite, Gaussian noise, the transfer function of the linear, physically realizable optimum predetection filter having infinite memory time is expressed as

$$H(j\omega) = \frac{1}{2\pi N_+(j\omega)} \int_0^\infty e^{-j\omega t}\, dt \int_0^\infty \frac{S^*(j\omega')e^{j\omega'(t-t_o)}}{N_+^*(j\omega')}\, d\omega' \tag{I-27}$$

where

ω' = variable of integration,

$*$ = complex conjugate,

$S(j\omega)$ = Fourier transform of the signal $s_i(t)$ at the input,

$N(\omega^2)$ = power spectrum of the noise $n_i(t)$ at the input,

$N_+(j\omega)$ = factor of $N(\omega^2)$ which, together with its conjugate, is analytic in the right half of the $j\omega$ plane, and thus, $N_+(j\omega)N_+^*(j\omega) = N(\omega^2)$.

In general, $N(\omega^2)$ is of the form

$$N(\omega^2) = \frac{a_o + a_1\omega^2 + \cdots + a_\ell\omega^{2\ell}}{b_o + b_1\omega^2 + \cdots + b_m\omega^{2m}} \tag{I-28}$$

where (m) and (ℓ) rarely exceed (3). If the input noise is assumed white, the transfer function obtained from Eq. (I-27) is in agreement with matched filter theory.

In the more practical case of finite memory, the situation is a little more complicated due to the requirement that $h(t)$, the impulse response, vanishes not only for $t < 0$ but also for

$t > T$, where (T) is a specified constant. The impulsive response of the optimum filter is found to be the sum of the impulse response for the infinite memory case plus three summations. The summations involve impulse functions of various orders, arising from discontinuities of $h'(t)$ and its derivatives at $t = 0$ and $t = T$ and the general solution of the differential equation $A(-p^2)h(t) = 0$ where $A(\omega^2)$ is the numerator of $N(\omega^2)$.

It is found that Eq. (I-26) is minimized if the impulse response satisfies the integral equation:

$$\int_0^T h(\tau)\, \psi_n(t-\tau)d\tau = s_i(t_o - t) \qquad \text{for} \quad 0 \leq t \leq T \tag{I-29}$$

where $\psi_n(\tau)$ is the correlation function of the noise component of the input to the filter. When $T = \infty$, Eq. (I-29) reduces to the Wiener-Hopf equation which is encountered in Wiener's theory of prediction. If the impulse response for finite observation time is the solution of Eq. (I-29), the mean-square value of the noise output of the optimum filter, σ^2, is numerically equal to the signal output at $t = t_o$, i.e.,

$$\sigma^2 = \int_0^T h(t)\, s_i(t_o - t)dt = s_o(t_o)\,. \tag{I-30}$$

Using Eqs. (I-25) and (I-30), the signal-to-noise ratio (ρ) at the output of the optimum filter is

$$\rho_{max} = \frac{|s_o(t_o)|^2}{\sigma^2} = s_o(t_o)\,. \tag{I-31}$$

It is important to note that the criterion used for both the matched filter and the Z-R optimum predetection filter is of the mean-square-error type. There are many criteria that may be used to evaluate optimum performance but the mse is chosen primarily for its accessibility to analytic manipulations. The main advantage of the above methods for optimum filter design is that they require relatively little statistical information about the noise and are thus less critically dependent upon the time and space stability of the signal and noise characteristics. A priori knowledge of the power spectrum or the correlation function of the noise is usually sufficient.

5. FILTERING IN AN IMPULSIVE NOISE BACKGROUND

An important class of interference which presents a different type of filtering problem is impulsive noise. Its distinguishing feature is that the energy occurs spasmodically, rather than continuously. Impulsive noise ordinarily has a wider spectral energy distribution than the signal. In underwater acoustics, impulsive interference sources may consist of explosives, earthquakes, or mechanical impacts generated at or near the receivers. Electromagnetic impulsive sources may consist of lightning discharges, or automotive and aircraft ignition. Statistical distributions of impulsive noise may be non-Gaussian and consequently the analyses of filtering problems previously described are not applicable. In the previous problems, the effectiveness of filtering operations was determined by comparison of the interference distributions with and without the signal being present. It was indicated that when the noise is white, and Gaussian the desired filtering operations can be determined — with an important advantage being that the only a priori information required is the mean power spectrum (or autocorrelation function), and the sole constraint being that the desired and undesired source are not correlated. Knowledge that the interference is Gaussian permits statistical predictions of performance — for example, for such distributions, the instantaneous value exceeds 3.09 × rms value for only 0.2 percent of the time.

Although for non-Gaussian distributions it is not possible to derive detailed analyses from knowledge of the spectrum, or autocorrelation function alone, it is nevertheless possible to establish upper bounds. If nothing is known of the distributions, then in accordance with a theorem by Tchebycheff, the portion falling outside (T) times the rms value must be less than $(1/T^2)$. Consequently, in order to obtain equivalent probability of "performance," it would be necessary to set $T = 22.36$ for an unknown distribution (in comparison to 3.09 for a Gaussian distribution) with $1/T^2 = 2$ in 1000. This type of operation is of course not effective, since it is based on using a rather primitive a priori specification — namely, only that the interference is non-Gaussian. If in addition, other information is available, then effective nonlinear filtering may be employed. Non-Gaussian interference can be analytically distinguished from Gaussian in terms of the concept of entropy power. For impulsive interference the entropy power is low — lower than its real power — and although it may have the same power spectrum as random noise it differs from random noise in having specifiable phase-relationships; its specific characteristic is a large amplitude which lasts for a short time.

The effectiveness of filtering is always based on the simultaneous operation on both the desired and undesired signal and consequently, it is not just the comparison of impulsive interference with Gaussian interference that is important. It is necessary to compare the structure of the desired signal with that of the undesired. In some instances, nonlinear "filtering" may be usefully applied to improve the response of the system in the absence of noise — operating on known characteristics of the signal which may be specified in terms other than its power spectrum. Examples of this are found in television where the picture sharpness may be increased by nonlinear filtering — apart from noise considerations.

One of the oldest impulsive noise-suppression techniques involves the use of an amplifier which has considerably greater bandwidth than that needed for the signal alone. The combined effects of the signal and interference will saturate the nonlinear circuits — since the clip-level is set just above the maximum expected value of the signal envelope. This operation will remove the peaks of the noise spikes. When the duration of the noise spike is short compared to the rate at which the signal envelope is changing, the remaining interference energy will be outside the signal band and may be removed by proper filtering.

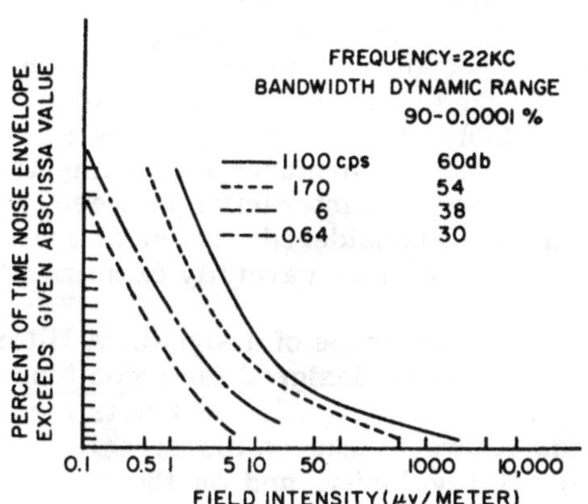

Figure I-7 - Amplitude distributions of atmospheric noise envelopes as functions of receiver bandwidth

It is important to recognize that the distribution of non-Gaussian interference is dependent on the time for which the noise is integrated before the variation of its envelope is determined. When the integration time is long (for example, when narrow band filters are used) the distribution approaches Gaussian — although for some types of interference it may be necessary to use very long integration times. Figure (I-7) illustrates the amplitude distributions of atmospheric noise envelopes as a function of receiver bandwidth. Figure (I-8) compares a type of impulsive noise with Gaussian, and in terms of the Central Limit Theorem, it is possible to convert a non-Gaussian into a Gaussian process by increasing the integration time. Based on this factor, it is possible in some problems to minimize the adverse effects of impulsive noise. Assume that the interference consists of a succession of transients whose duration is approximately equal to the reciprocal of the system bandwidth. If these impulses are passed through a linear filter whose impulse response has a very large TW product, then the transient response will last for a longer time than the channel response. A filter transient response of this nature is indicated by Figure (I-9). Energy delivered in the original impulse will be "smeared" over the time interval T_F, and consequently the peak amplitude will be reduced by the ratio of the channel response to the smearing time. The smoothing may be continued to the point at which the length of the smearing becomes comparable to the average time interval between pulses.

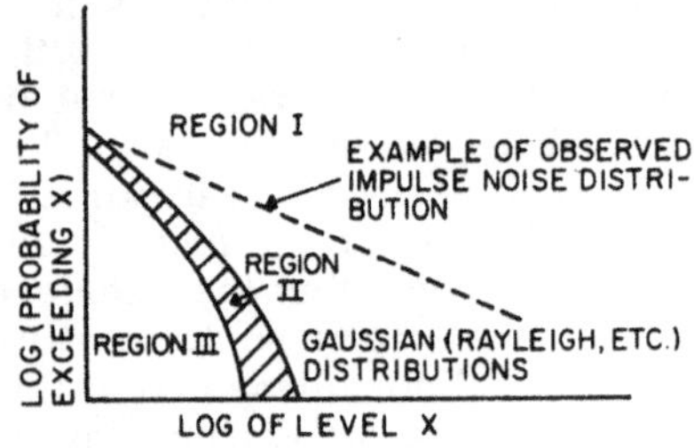

Figure I-8 - Noise probability distributions

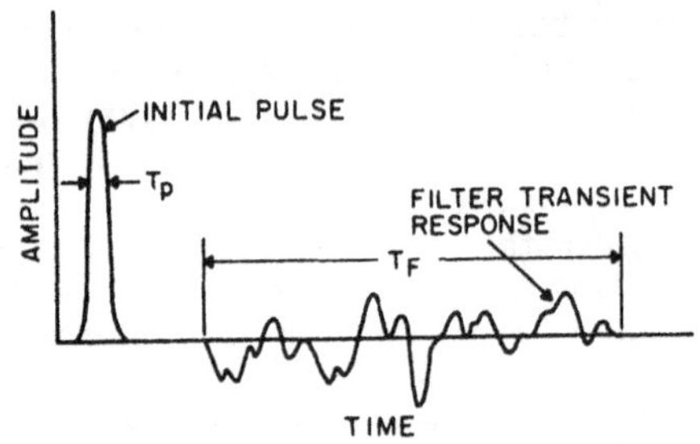

Figure I-9 - An example of a complicated impulse response

The smearing of the noise must be done in such a way that the signal itself is not badly degraded. If a total bandwidth of W is required in the channel and then, if tuned-circuit filters are used whose bandwidth is inversely dependent on the smearing time, then a total of $2T_FW$ channels would be required to accommodate the total bandwidth W. The requirement for multiple channel processing may be eliminated by using encoded transmissions wherein the transmission would be in terms of waveforms which differ from an impulse and whose duration would extend over the time interval T_F. As a result, at the receiver, the decoding process would consist of processing over the interval T_F, thereby automatically performing the smearing process on the interference.

6. EXTRACTION FILTER

INTRODUCTION

There are many factors to consider when "optimizing" a filter's performance. Optimization will depend on the purpose of the filter, the nature of the inputs, the criterion employed for evaluating performance, and component tolerances. In the preceding discussions, the problems involved detecting the presence or absence of a signal masked by a noise background. When the spectra of the signal and noise do not overlap appreciably or are different in their time structure, then various filtering methods may be used. As the performance requirements become more severe, analysis becomes more complex. Effects such as interaction between spectra, rate of change of spectra, and finite memory time must be considered. Physical realizability and significance of criteria employed in design needed to be more carefully examined.

An important problem involves preserving or extracting the waveshape of a signal. A filter which performs this operation is called an extraction filter and may be designed for smoothing or predicting or may combine both operations. A smoothing and predicting filter extracts the wanted signal from a signal plus noise complex and yields future values of the signal. Physical prediction depends on the process having statistical regularity, and on the existence of correlations between future values of the signal and past values of the known data. If the prediction is accomplished by a linear operation, then the only type of correlation that can be used is linear correlation. This has the disadvantage of not making complete use of possible relationships contained, for example, in higher moments. It does have the advantage of simplifying the analysis and facilitating synthesis of the optimum filter. Often, a linear prediction is the best that can be done though it may be inadequate. The application of correlation and spectral analysis to the design of linear systems for statistical smoothing and prediction was first proposed by Wiener.

LINEAR LEAST SQUARE SMOOTHING AND PREDICTION

There are three main assumptions upon which the application of the Wiener theory depends. These assumptions are:

1. The time series represented by the signal $s(t)$ and the noise $n(t)$ are stationary. This is to insure that statistical regularities observed in the past will continue in the future (the statistical properties do not change with time).

2. The prediction and smoothing is obtained by a linear operation, that is, with a linear, physically realizable filter. Linearity encourages generalization of theory to include a wide class of signals and realizability requires the results be practical.

3. The measure of effectiveness of the filter is the mean-square difference between the actual output and the desired output. This is an ensemble average which uses the statistics of the amplitudes of the different frequency components of the signal and noise.

The Wiener filter performs linear least square prediction and smoothing of a stationary time series. If $h_e(t)$ is the impulse response of an ideal linear filter whose output is $e(t)$, then the following characteristics are desired for an input signal, $s(t)$:

(a) Ideal prediction

$$e(t) = s(t+\alpha)$$

$$h_e(t) = \delta(t+\alpha)$$

$$H_e(j\omega) = e^{j\omega\alpha}$$

(b) Ideal smoothing

$$e(t) = s(t)$$

$$h_e(t) = \delta(t)$$

$$H_e(j\omega) = 1$$

where $H_e(j\omega)$ is the transfer function of the ideal filter and (α) is positive, signifying a time advance. As a result of filtering linearly,

$$e(t) = \int_{-\infty}^{\infty} h_e(\beta)\ s(t-\beta)d\beta \tag{I-32}$$

which is not necessarily physically realizable. The above also indicates that if the smoothing problem is solved, it may be easily extended to include prediction by introducing a time advance (α) in the time solution or a continuous, linear advance in phase $e^{j\omega\alpha}$ in the frequency solution.

Let the signal plus noise be denoted by $x(t) = s(t) + n(t)$ and defined for $-\infty \leq t \leq T$, where (T) is the present time. The problem then reduces to finding the best mean-square estimate of $e(T)$ that is generated by a physically realizable linear operation on $x(t)$. The procedure is shown in Figure (I-10) where $\hat{e}(T)$ is the estimate of $e(t)$ at time (T), and $h(t)$ and $H(j\omega)$ are the impulse response and transfer function of the linear filter, respectively. Applying the mean-square error criterion to the output of Figure (I-10),

$$\text{mse} = E\left[\mathcal{E}^2(T)\right] = E\left[e(T) - \hat{e}(T)\right]^2 \tag{I-33}$$

where $E[\ ]$ refers to an ensemble average of all possible signal and noise functions with each weighted according to its probability of occurrence. Expanding the right side of Eq. (I-33),

$$\text{mse} = E\left[e^2(T)\right] - 2E\left[e(T)\ \hat{e}(T)\right] + E\left[\hat{e}^2(T)\right]. \tag{I-34}$$

The signal and noise are random processes assumed to be statistically independent, stationary, have zero means, and autocorrelation functions $\psi_s(\tau)$ and $\psi_n(\tau)$, respectively. With the aid of the convolution theorem, Eq. (I-34) may be rewritten as

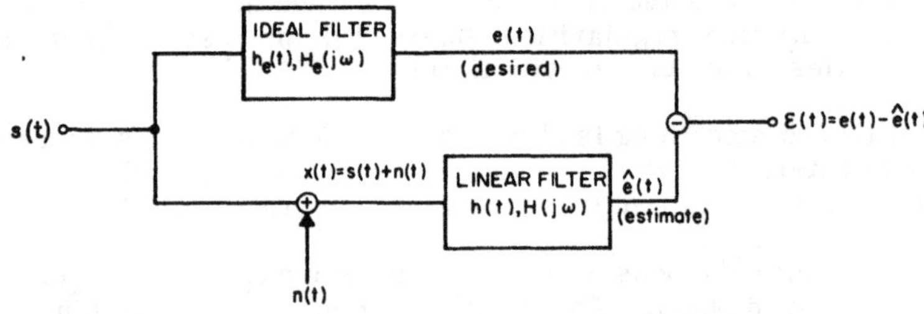

Figure I-10 - Procedure for estimating the desired output with a linear, physically realizable filter

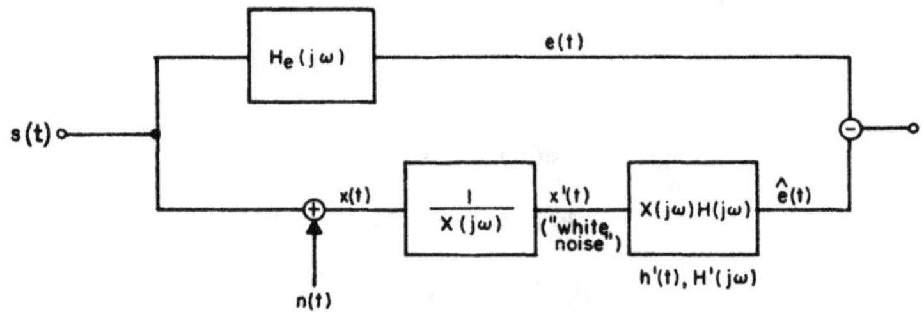

Figure I-11 - Procedure using spectrum-shaping technique for obtaining an optimum smoothing and prediction filter, H(jω)

$$\text{mse} = \psi_e(0) - 2\int_0^\infty h(\beta)\,\psi_{xe}(\beta)d\beta + \int_0^\infty\int_0^\infty h(\beta_1)\,h(\beta_2)\,\psi_x(\beta_1-\beta_2)\,d\beta_1\,d\beta_2\,, \tag{I-35}$$

where $\psi_e(0)$ is the average power in $e(t)$, $\psi_{xe}(\tau)$ is the crosscorrelation function between $e(t)$ and $x(t)$, and $\psi_x(\tau)$ is the autocorrelation function of $x(t)$. The problem now is to find a function $h(t)$ which minimizes the integrals in Eq. (I-35). This suggests using the calculus of variations. Applying this technique to Eq. (I-35), the mean-square discrepancy between the desired and actual output will be minimized if the impulse response of the optimum filter $h(t)$ satisfies the following relationship:

$$\int_0^\infty h(\beta)\,\psi_x(t-\beta)d\beta = \psi_{xe}(t) \qquad \text{for all } t \geq 0. \tag{I-36}$$

This is known as the Wiener-Hopf equation.

If $x(t)$ is reduced to white noise, then the Wiener-Hopf equation may be easily solved. In order to make use of this fact, a spectrum-shaping technique is introduced. If the amplitude-phase spectrum of the input is denoted by $X(j\omega)$, then the new procedure for obtaining an optimum filter is shown in Figure (I-11) where $x'(t)$ represents white noise. The Wiener-Hopf equation is then expressed as

$$\int_0^\infty h'(\beta)\,\psi_{x'}(t-\beta)d\beta = \psi_{x'e}(t) \qquad \text{for all } t \geq 0. \tag{I-37}$$

Since the autocorrelation function of white noise is an impulse function, Eq. (I-37) reduces to

$$h'(t) = \psi_{x'e}(t) \qquad t \geq 0$$
$$= 0 \qquad t < 0 \tag{I-38}$$

It is often desirable to work in the frequency domain when using spectrum shaping techniques. For example, it is convenient to determine the cross-power spectra between $x'(t)$ and $e(t)$. This is

$$W_{x'e}(\omega) = \frac{H_e(j\omega)\, W_s(\omega)}{X(-j\omega)} \tag{I-39}$$

where $W_s(\omega)$ is the power spectrum of the signal and $X(-j\omega)$ is a factor of the power spectrum of the input $W_x(\omega)$ such that $W_x(\omega) = X(j\omega)\, X(-j\omega)$; $X(s)$ has all poles and zeros in the LHP. Equation (I-39) may then be expanded as

$$W_{x'e}(\omega) = W_\ell(\omega) + W_r(\omega) \tag{I-40}$$

where $W_\ell(\omega)$ corresponds to LHP poles and $W_r(\omega)$ corresponds to RHP poles. Taking the Fourier transform of both sides,

$$\psi_{x'e}(t) = h_\ell(t) + h_r(t) \tag{I-41}$$

and has the following constraints:

$$h_\ell(t) = 0 \qquad \text{for } t < 0$$
$$h_r(t) = 0 \qquad \text{for } t > 0 \tag{I-42}$$

Comparing Eqs. (I-41) and (I-42) to Eq. (I-38), it is seen that only the poles in the LHP need be considered to insure physical realizability.

The solution of the linear least square smoothing and prediction problem may be summarized by the following steps:

1. Reduce the input $x(t)$ to white noise. $x'(t)$

2. Expand $W_{x'e}(\omega) = H_e(j\omega)\, W_s(\omega)/X(-j\omega)$ in partial fractions. Let $h'(t) = \Sigma$ terms corresponding to LHP poles.

3. Compute $H'(j\omega)$ by taking the Fourier transform, $\mathfrak{F}\{\ \}$, of $h'(t)$

$$H'(j\omega) = \mathfrak{F}\{h'(t)\}.$$

4. Obtain the optimum filter whose transfer function is

$$H(j\omega) = H'(j\omega)\, \frac{1}{X(j\omega)}$$

which is physically realizable.

SIGNIFICANCE OF MEAN-SQUARE ERROR CRITERION

The design formulas for the Wiener optimum filter depend only on the power spectrum of the signal and noise. Consequently, it may seem necessary only to consider the statistical distribution of their amplitudes and not of their phases. However, the Wiener filter filters on the basis of waveshape and not just the spectrum so that relative phases of the signal and noise must be considered. Because it is required that the prediction be a linear operation, a compromise is forced upon the design — that is, instead of using relative phase information properly, the compromise essentially averages over the relative phases of the various components of the signal. Two different types of signal with the same spectrum may produce the same optimum filter and the same mean-square error; however, the effectiveness of the filter may not be the same for both signal types. Phase information can improve prediction and smoothing along with the use of nonlinear elements (or possibly synchronous linear time-varying elements). Nonlinear prediction theory would introduce phase correlation as an additional parameter to consider.

An important assumption in prediction theory is the mean-square-error criterion. This primarily minimizes the effect of the large errors without giving much weighting to the small errors. When predicting, it is not always the size of the error that is important, for example, the number of errors may also be of significance. In such problems, the small minor errors neglected by the mean-square-error criterion will have equal importance in describing the performance of the filter as the large errors. Other criteria may be used, such as maximizing a conditional probability which would treat all errors equally but requires a complete statistical knowledge of inputs. Another criterion may be that of minimizing the probability of exceeding a certain threshold. However, when the distribution of future events is Gaussian, it does not matter what criterion is used since the most probable event is the one for which the mean-square error is least.

A least mean-square prediction is one which selects the mean value of the distribution of possible future responses since this is the point about which the mean-square error is least. When predicting, it is desired to select the value of the signal which corresponds to the maximum probability of the future distribution. However, the position of the mean is usually a nonlinear function of the past history of the input and thus, does not necessarily correspond to maximum probability. Only if the future responses are distributed according to a Gaussian distribution does the best linear predictor select the center of the distribution for the predicted value. Thus, in the Gaussian case a nonlinear operation or non-mean-square-error criterion would not provide additional prediction accuracy but would only increase the complexity of instrumentation and computation.

REFERENCES

I-1. R. Saal and E. Ulbrich, "On the Design of Filters by Synthesis," I.R.E. Trans., Vol. CT-5, No. 4, December 1958, pp. 284-327.

I-2. N. Wiener, "Extrapolation, Interpolation, and Smoothing of Stationary Time Series," New York, John Wiley and Sons, Inc., 1949.

I-3. L. A. Zadeh and J. R. Ragazzini, "An Extension of Wiener's Theory of Prediction," Journal Applied Physics, Vol. 21, pp. 645-655, July 1950.

I-4. R. C. Booton, "An Optimization Theory for Time-Varying Linear Systems with Nonstationary Statistical Inputs," Proc. I.R.E., Vol. 40, pp. 977-981, 1952.

I-5. L. A. Zadeh, "Optimum Non-linear Filters," Journal Applied Physics, Vol. 24, pp. 396-404, 1953.

I-6. D. O. North, "Analysis of the Factors which Determine Signal/Noise Discrimination in Radar," Report PTR-6C, R.C.A. Laboratories, 1943.

I-7. L. A. Zadeh and J. R. Ragazzini, "Optimum Filters for the Detection of Signals in Noise," Proc. I.R.E., Vol. 40, pp. 1223-1231, 1952.

I-8. Y. W. Lee, T. P. Cheatham, Jr., and J. B. Wiesner, "Application of Correlation Analysis to the Detection of Periodic Signals in Noise," Proc. I.R.E., Vol. 38, pp. 1165-1171, 1950.

I-9. J. V. Harrington, "Signal-to-Noise Improvement Through Integration In a Storage Tube," Proc. I.R.E., Vol. 38, pp. 1197-1203, October 1950.

I-10. J. V. Harrington, "An Analysis of the Detection of Repeated Signals in Noise by Binary Integration," I.R.E. Trans., Vol. IT-1, No. 1, pp. 1-9, March 1955.

I-11. G. L. Turin, "An Introduction to Matched Filters," I.R.E. Trans., Vol. IT-6, No. 3, pp. 311-330, June 1960.

I-12. G. L. Turin, "On the Estimation in the Presence of Noise of the Impulse Response of a Random, Linear Filter," I.R.E. Trans., Vol. IT-3, pp. 5-10, March 1957.

I-13. G. L. Turin, "Error Probabilities for Binary Symmetric Ideal Reception Through Non-selective Slow Fading and Noise," I.R.E. Proc., Vol. 46, pp. 1603-1619, September 1958.

I-14. R. M. Lerner, "A Matched Filter Detection System for Complicated Doppler Shifted Signals," I.R.E. Trans., Vol. IT-6, No. 3, pp. 373-385, June 1960.

I-15. R. C. Titsworth, "Coherent Detection by Quasi-Orthogonal Square-Wave Pulse Functions," I.R.E. Trans., Vol. IT-6, No. 3, pp. 410-411, June 1960.

I-16. D. Middleton, "On New Classes of Matched Filters and Generalizations of the Matched Filter Concept," I.R.E. Trans., Vol. IT-6, pp. 349-360, June 1960.

I-17. H. W. Bode and C. E. Shannon, "A Simplified Derivation of Linear Least Square Smoothing and Prediction Theory," I.R.E. Proc., pp. 417-425, April 1950.

I-18. S. Goldman, "Information Theory and Radar," Naval Research Reviews, pp. 8-11, January 1959.

I-19. E. Baghdady, Ed., "Lectures on Signal Analysis," McGraw-Hill Book Co., Inc., New York, N.Y., Chapter 10, 1961.

I-20. S. Darlington, "Linear, Least-Squares Smoothing and Prediction, With Applications," Bell Sys. Tech. J., Vol. 37, pp. 1221-1294, September 1958.

I-21. S. Darlington, "Nonstationary Smoothing and Prediction Using Network Theory Concepts," Trans. 1959 International Symp. on Circuit and Information Theory.

I-22. P. M. Woodward, "Probability and Information Theory With Applications to Radar," McGraw-Hill Book Co., Inc., New York, N.Y., 1953.

I-23. W. W. Petersen, T. G. Birdsall, and W. C. Fox, "Theory of Signal Detectability," I.R.E. Trans., Vol. IT-4, pp. 171-212, September 1954.

I-24. W. M. Siebert, "A Radar Detection Philosophy," I.R.E. Trans., Vol. IT-2, pp. 204-221, September 1956.

I-25. M. Zakai, "On a Property of Wiener Filters," I.R.E. Trans., Vol. IT-5, No. 1, pp. 15-17, March 1959.

I-26. D. Middleton and D. Van Meter, "Detection and Extraction of Signals in Noise from the Point of View of Statistical Decision Theory, Parts I and II," J. Soc. Ind. Appl. Math., Vol. 3, pp. 192-253, December 1955, and Vol. 4, pp. 86-119, June 1956.

I-27. D. Middleton, "An Introduction to Statistical Communication Theory," McGraw-Hill Book Co., Inc., New York, N.Y., 1958.

I-28. J. H. Park, Jr., "Statistical Estimation of Normalized Linear Signal Parameters," Rad. Lab., The Johns Hopkins University, Balt., Md., Tech. Rept. AF-72, December 1959.

I-29. E. M. Glaser, "Signal Detection by Adaptive Filters," Rad. Lab., The Johns Hopkins University, Balt., Md., Tech. Rept. AF-75, April 1960.

I-30. P. Elias, A. Gill, R. Price, N. Abramson, P. Swerling, and L. Zadeh, "Progress in Information Theory in the U.S.A., 1957-1960," I.R.E. Trans., Vol. IT-3, pp. 128-144, July 1961.

I-31. S. F. George and A. S. Zamanakos, "Comb Filters for Pulsed Radar Use," I.R.E. Proc., Vol. 42, No. 7, July 1954, pp. 1159-1165.

I-32. J. Galejs, "Enhancement of Pulse Train Signals by Comb Filters," I.R.E. Trans., Vol. IT-4, No. 3, September 1958, pp. 114-125.

J. SPACE-TIME ASPECTS

1. INTRODUCTION

Previous sections have indicated the progressive evolution of the complexity of descriptions of circuit and spatial elements. Initially, the primary concern involved structural considerations. The effects of boundaries and of errors resulting from approximation were then discussed, along with methods for assessing performance when various types of signal and interference are present, and when multiple functions are to be performed simultaneously — for example, determining time of arrival and frequency content. Although not developed in detail, it was indicated that informational criteria and decision theory offered advantages in comparison with the use of signal-to-noise measures. In particular, decision theory concepts are readily extended to include space, and space-time relationships. Relationships among the functions being performed, use of a priori information, and the effects of uncertainties are to be included in the actual formulation of the problem.

Situations which characterize many important informational processes are functions of both space and time. Such problems increase in complexity not only because of the added number of dimensions but also because the physical interpretation's become considerably more difficult. Space-time processes may require the solutions of partial differential equations, for which, in contrast to ordinary differential equations, generalized methods for obtaining solutions are not available. Complete solutions necessitate the use of specific boundary conditions and determining the physical validity of the assumptions made regarding boundary conditions.

When only the space domain was considered previously, an analogy to a time-and-band-limited waveform was a spatial element of finite extent and whose pattern was restricted to the accessible portion of the field, that is, the region of real angles. Space-time processes involve multiple boundaries imposed in the space and time domains. The analyses of such problems should be sufficiently complete to establish the primary informational features which include not only the structural aspects, but the full range of detail which is observable within the structural elements. Observations may not be made of infinite detail since fluctuations ultimately impose a limit when finite observations are involved. Even in the absence of fluctuations, when the process includes multiple operations which are to be performed simultaneously, indeterminacies may exist which constrain the detail which can be observed. The analysis should not only establish these features but should also provide comparison of the effectiveness or assessments of relative costs. This may be done, for example, by the requirement that the least number of independent parameters be employed, or that they be used most economically.

It will become evident that apart from the greater complexity, there still remains a conceptual correspondence among the methods which may be employed with space-time process, and those which have been earlier derived for simple elements and processes. As examples, useful quantitative measures may be obtained from space-time sampling, space-time correlation functions, and space-time ambiguity functions. Probabilistic, deterministic and combined probabilistic and deterministic measures may be required and for reasons similar to those given in the earlier sections. Fourier methods were found to be useful in space and time problems, and illustrated some of the basic concepts of transformation and representation philosophy. A correspondingly important role is played by "waves" — that is, responses which exist and are propagated over spatial coordinates. In all cases it is necessary to keep clearly in mind the limitations of the methods. In particular, although a number of physical problems in which the wave equation may be applied have common mathematical bases, the physical verification, that is, agreement with observation is the ultimate test of the validity of the analysis.

Complete and detailed descriptions of space-time processes even for relatively simple acoustical problems would transcend the scope of this survey. Such descriptions would require consideration of energy and information transfer taking into account dissipative and dispersive properties of the medium. Descriptions of the physical features of the environment are often made, in a sense, independently of the method of observation, and of the interrelationships and interactions of the measuring instrumentation with the process being described and observed. Additionally, the effects of assumptions made in order to simplify the mathematical analyses — which involve concepts as plane waves, "fields," "point" sources, coherency and incoherency — are not always evident in the descriptions of the problems. An important factor to be illustrated is the requirement for describing the interrelationships completely, by including the limitations of the observation process, and the effects of assumptions and constraints in the formulation and analysis of the problem.

2. ACOUSTIC WAVE EQUATION

If a fluid is in a state of equilibrium and the pressure in a certain region is changed from its equilibrium value, the fluid will immediately produce forces which tend toward restoring the equilibrium value. Vibrations result, which are propagated as waves through the fluid. The restoring forces are attributed to the elasticity of the medium and the wave propagation due to the inertia of the displaced particle, permitting a transfer of momentum to adjoining particles.

Assume an isotropic, homogeneous, perfectly elastic and unbounded medium, supporting acoustic waves of relatively small amplitude. The velocity vector $\mathbf{V}$ of a particle at point (x, y, z) at time (t) may be represented as the gradient of a scalar velocity potential function Ψ, as indicated in section H-7, i.e.,

$$\mathbf{V} = \nabla\Psi(x, y, z, t) . \qquad \text{(H-5)}$$

Applying the principle of continuity and the additional assumptions that the fractional change of the instantaneous density from its mean value is small compared with unity, and the resolved vector velocities change slowly with respect to their corresponding ordinates, Eq. (H-5) becomes

$$\nabla^2\Psi - \frac{1}{c^2}\frac{\partial^2\Psi}{\partial t^2} = 0 \qquad \text{(J-1)}$$

where ∇^2 is the Laplacian operator

$$\nabla^2 = \frac{\partial^2}{\partial x^2} + \frac{\partial^2}{\partial y^2} + \frac{\partial^2}{\partial z^2} \qquad \text{(J-2)}$$

and (c) is the speed of propagation through the medium. Equation (J-1) is the three-dimensional form of the acoustic wave equation and gives the relationship between the time and space derivatives of the velocity potential in the fluid (liquid or gas) satisfying the above conditions.

Physically, the acoustic wave equation gives a dynamical description of a physical process relating various temporal and spatial rates of change. Mathematically, it is a linear homogeneous partial differential equation of the second order, and may be written as

$$L[\Psi] = 0 \qquad \text{(J-3)}$$

where L is the linear differential operator or Lorentzian operator,

$$L = \nabla^2 - \frac{1}{c^2}\frac{\partial^2}{\partial t^2} \qquad \text{(J-4)}$$

satisfying the principle of superposition. A general solution of a partial differential equation like (J-3) always contains arbitrary constants and functions. These can only be evaluated by considering the specific conditions of individual problems. Descriptions of space-time processes necessitates a consideration of boundary and initial conditions.

Boundary conditions are fixed by the geometry of the medium itself while initial conditions are concerned with the disturbance which causes the acoustic wave to be propagated. An initial condition may specify a distribution (of pressure, velocity, particle displacement, etc.) at a certain instant of time over the whole fluid, or some acoustic variable as a function of time at a fixed point. The wave equation, (J-3), says nothing of the disturbance or mechanism generating the wave and applies only to the region excluding sources. Mathematically, this implies that

$$\Psi^{(\alpha)}(o,o,o,t) = 0 \qquad , \alpha = 0,1 . \tag{J-5}$$

Such conditions are classified as homogeneous. Homogeneous boundary or initial conditions usually consist of the relations between the values assumed by the desired function and its derivatives on some closed surface of the domain in question. When an excitation is applied, the problem changes from solving a homogeneous wave equation with homogeneous boundary (or initial) conditions to one having nonhomogeneous boundary (or initial) conditions. A very useful concept in boundary-value problems is that, in general, homogeneous differential equations with nonhomogeneous boundary conditions are essentially equivalent to nonhomogeneous differential equations with homogeneous boundary conditions. This entails basic assumptions as to continuity and differentiability.

The general solution of the wave equation depends on the symmetry of the propagating wave which is determined by the source geometry. For example, if there is symmetry with respect to an infinite plane, energy will be propagated in one direction only, normal to the plane. For one-dimensional (x) motion the wave equation becomes

$$\frac{\partial^2\Psi}{\partial x^2} - \frac{1}{c^2}\frac{\partial^2\Psi}{\partial t^2} = 0 . \tag{J-6}$$

The general solution of Eq. (J-6) is of the form

$$\Psi(x,t) = f_1(x-ct) + f_2(x+ct) \tag{J-7}$$

where f_1 and f_2 are arbitrary functions depending only on the initial and boundary conditions. The waves (J-7) are called plane waves and have the characteristic property that the acoustic pressures, particle displacements, etc., have common phases and amplitudes at all points on any plane perpendicular to the direction of wave propagation.

Sound at large distances from an actual source resembles the sound from a point source more closely than it does that from an infinite plane. Consider an acoustic wave which is propagated in three dimensions and is symmetrical with respect to a point source generating it. The pressure or velocity potential in the surrounding medium will be a function of the distance (r) from the point source only and of the time (t). The Laplacian operator reduces to

$$\frac{\partial^2}{\partial r^2} + \frac{2}{r}\frac{\partial}{\partial r}$$

in spherical coordinates and the wave equation may be expressed as

$$\frac{\partial^2\Psi}{\partial r^2} + \frac{2}{r}\frac{\partial\Psi}{\partial r} - \frac{1}{c^2}\frac{\partial^2\Psi}{\partial t^2} = 0 \tag{J-8}$$

or

$$\frac{\partial^2(r\Psi)}{\partial r^2} - \frac{1}{c^2}\frac{\partial^2(r\Psi)}{\partial t^2} = 0 . \tag{J-9}$$

Comparing Eq. (J-9) with Eq. (J-6),

$$\Psi(r,t) = \frac{1}{r} f_1(r - ct) + \frac{1}{r} f_2(r + ct) . \tag{J-10}$$

The first term of Eq. (J-10) represents a spherical wave diverging from the origin of the coordinates and the second term represents a similar wave converging on the origin. The latter is of little importance in acoustics. In contrast with plane waves, the particle velocity of acoustic waves possessing spherical symmetry is generally not in phase with the pressure. However, the wave fronts of many types of divergent waves in a homogeneous medium assume the characteristics of plane waves as they proceed to great distances from the generators.

Another important type of acoustic disturbance is that propagated in three dimensions and is symmetrical about an axis such as that of a cylindrical source. The Laplacian operator reduces to

$$\frac{\partial^2}{\partial r^2} + \frac{1}{r}\frac{\partial}{\partial r}$$

in cylindrical coordinates where (r) is now the distance measured from the polar axis. The wave equation then becomes

$$\frac{\partial^2 \Psi}{\partial r^2} + \frac{1}{r}\frac{\partial \Psi}{\partial r} - \frac{1}{c^2}\frac{\partial^2 \Psi}{\partial t^2} = 0 . \tag{J-11}$$

The solution to Eq. (J-11) represents a cylindrical wave and is given in terms of Bessel functions of the first and second kind of zero order. Both functions are similar for large values of (r), where they are approximately sinusoidal with amplitudes varying inversely as the square root of (r). In the vicinity of the origin of the coordinate system, Bessel functions of the first kind are finite while those of the second kind are infinite. Bessel functions occur in the theory of cylindrical and spherical waves just as sinusoidal functions appear in the theory of plane waves.

Thus far, the wave equation has been discussed without considering the type of source distribution or initial condition. If the displacements caused by the excitation produce restoring forces proportional to the displacement, then the initial acoustic disturbance will have the form of a harmonic vibration, producing a pure tone. For example, assume that as an initial condition, the velocity potential (or pressure) is specified only at the source for the time interval between $t = 0$ and $t = T$ and is zero for $t < 0$ and $t > T$. Assume also that the waves are propagated in the positive x direction only. The harmonic solution for an infinite plane source is, from Eq. (J-7),

$$\Psi(x,t) = A \cos 2\pi f(t - x/c) \tag{J-12}$$

where (A) is the maximum amplitude of the disturbance and (f) is its frequency. The region of disturbance is always of width cT and is propagated with the speed (c). For a point source, if the initial disturbance is confined to a spherical shell of infinitesimal thickness at a distance $r = r_o$ from the origin (this is necessary due to the infinite value predicted for Ψ at the origin in Eq. (J-10)) between the times $t = 0$ and $t = T$, then if (r_o) is nearly zero, the velocity potential at a distance (r) from the harmonic source and time (t) is given by

$$\Psi(r,t) = \frac{A \cos 2\pi f(t - r/c)}{r} . \tag{J-13}$$

Clearly, the maximum pressure change at the distance (r) is given by A/r, decreasing as (r) increases.

In practice, as a first approximation, restoring forces may be considered to be proportional to the particle displacements for restricted portions of the medium. With the aid of Fourier analysis, the harmonic vibration may be the building block for the more complex solutions of the wave equation. This is called normal mode theory. A normal-mode solution is one where the motion or vibration is described by a set of discrete frequencies whose amplitudes vary in a sinusoidal manner. It is extremely useful in problems where standing wave patterns are produced and fixed boundary conditions have to be met. The general expression for the velocity potential may then be expressed as a finite sum of the form

$$\Psi(x, y, z, t) = \sum_n A_n(x, y, z) \cos 2\pi nft . \qquad \text{(J-14)}$$

The important feature of the representation is that the space and time dependence occurs as the product of separate factors. If Eq. (J-14) is substituted into the wave equation (J-1), then $A(x, y, z)$ satisfies the partial differential equation

$$\nabla^2 A(x, y, z) + \left(\frac{2\pi n}{\lambda}\right)^2 A(x, y, z) = 0 \qquad \text{(J-15)}$$

which is time independent. The function $A(x, y, z)$ is referred to as an eigenvibration or "proper" function and the parameter $k_n = (2\pi n/\lambda)$ is its eigenvalue or "proper" value. Normal mode solutions with the associated concept of orthogonality have been discussed in section F. Note that Eq. (J-14) is the solution to a fixed boundary problem when the initial pressure disturbance can be expressed as a finite sum of eigenvibrations. The utility of normal mode theory is that it provides solutions which lend themselves to practical problems. However, this is done at the expense of excluding the effect of interactions between the space and time variables, which may often prove to be the primary consideration.

In order to produce a disturbance in a medium which propagates as a wave — energy must be transferred to the medium. Energy propagation is an important aspect of wave motion. For a material medium the product of the stress and the displacement velocity at a point in the medium gives the rate at which energy is being communicated, per unit area of wavefront, to the medium by the wave at this point. This rate varies in time and space. For a harmonic, progressive wave it is necessary to take an average over time at a particular point in space and obtain the average flow of power per unit area of the wavefront. The average flow of power per unit area in the wave may be used to represent the intensity of the wave. In general, for a wide variety of waves, the power per unit area is proportional to the time derivative of a quantity representing the deformation with the space derivative of the same quantity.

Since an averaging process is involved in defining intensity, different times of averaging, particularly if short compared to the period of a harmonic wave, will give different intensities. All measuring devices require a finite time to respond to an excitation. An instrument which averages over a short period will indicate fluctuations which are not apparent for instruments having considerably longer response times. In comparing measurements with theory, the nature of the measuring device must be taken into account — and for space-time processes, in addition to the consideration given to the bandwidth, the spatial extent of the instrument within the medium must be taken into account.

An important aspect associated with energy considerations is that it is possible to derive the wave equation using techniques employed in mechanics for obtaining the equations of motion of a dynamical system of particles. Additionally, by using variational methods involving total energy, analysis of problems may be facilitated. Since the wave intensity expresses the average power transmission per unit area of wavefront in the wave, it may also be represented as the product of the wave velocity and the average total energy per unit volume in the portion of the medium being traversed by the wave — energy per unit volume is designated as the energy density. The average total energy density is the sum of the average kinetic energy density and the average potential energy density. For a plane harmonic progressive dilatational wave, the energy density is equally divided between kinetic and potential. For stationary waves, particularly when a number of modes may be present, the space-time average of the kinetic energy density is the sum of the average kinetic energy densities for the various modes. Here, too, the averaging time and the specific details of the measuring instrument play an important role in determining the existence and magnitude of fluctuations. For stationary waves, it is evident that energy density is a more meaningful measure than intensity.

Other representations of wave propagation may be made in terms of the motion of the wave fronts — that is, surfaces of constant amplitude and phase. Ray theory permits constructing curves which are normal at every point to the wavefronts, and which provide simplified means for determining some of the spatial propagation features, and for computing travel times. If the medium is characterized by a constant phase velocity, the rays are straight lines. In a stratified medium in which the wave velocity variation is normal to the layer, the rays are plane curves in planes normal to the layers, the curvature obeying Snell's law.

Although ray tracings provide a convenient and useful representation, quantitative measures of energy or intensity and of fluctuations, particularly when boundaries intervene, are not readily obtained.

3. SPACE-TIME RADIATION PATTERN

INTRODUCTION

The previous discussion involved solutions of the wave equation showing some of the effects of boundaries, source geometry, and the form of the excitation. The specific methods used to obtain the solution encompasses the nature of the boundaries. For example, the boundaries of the propagating medium and the source geometry influence the coordinate system to be used in order to facilitate separating the space and time variables. Excitation of the source establishes the initial conditions. The use of energy and intensity relationships was also briefly discussed, including the characteristics of measuring instrumentation. The examples illustrated the basic premise that generalized procedures for space-time processes are inherently more difficult to obtain, and that the assumptions made which facilitate mathematical analyses must be validated by physical interpretation.

Huygen's principle is of considerable importance in radiation problems. It states that if a source of radiation is completely enclosed by a hypothetical surface (S), then there is an equivalent source which when spread over (S) will give the same field outside (S) as the actual source. Thus, if the field is known at all points of the surface, the field at exterior points (M) can be determined by summing the contributions to the field at (M) from each elemental area (dA) of the surface, assuming each to be a secondary source of radiation. The equivalent source is not unique, that is, although different forms may give different internal fields, they will give the same external field.

Even when the properties of the medium are taken to be constant, complex initial conditions can introduce complexity into the analysis. When a source distribution is established in the region of interest, its stress effects must be added to the equation of motion. If Eq. (J-3) applies to a region which does not contain sources, then the acoustic wave equation for a region with a source distribution (Φ) may be written as

$$L[\Psi] = -\Phi . \tag{J-16}$$

The quantity (Φ) is the source strength per unit volume or the volume rate at which fluid is injected into the medium.

Consider a plane-surface radiator of finite extent whose active area is bounded by the lines $x = \pm X/2$ and $y = \pm Y/2$, where any elementary area of this source has the source strength $\Phi(x,y)\,dxdy$. The far-field pattern in the y-z plane may be obtained by determining the pattern of an equivalent line source along the y-axis having source strength distribution $F(y)$, where

$$F(y) = \int_{-X/2}^{X/2} \Phi(x,y)\,dx . \tag{J-17}$$

A convenient means of synthesizing the y-z plane and x-z plane pattern functions of a plane-surface source is by requiring their source strengths to be independent so that they may be treated separately, i.e.,

$$\Phi(x,y) = \Phi(x)\;\Phi(y) . \tag{J-18}$$

This effectively reduces the area source to a pair of crossed line sources. The pattern function in any other plane through the z-axis is found by an appropriate product of the x-y and y-z patterns. In synthesizing pattern functions, the distribution of source strength on the surface of the source is specified which implies specifying the relative velocity of vibration of the source.

GENERAL RADIATION PATTERN

Section G-12 indicated how the concept of the Green's function facilitated analysis of time-varying elements. In particular, Green's function permitted the solution of a nonhomogeneous differential equation describing the behavior of the element, having homogeneous initial conditions, to be written as an integral equation. Analogously, Green's function can be applied to solving acoustic wave equation for time-varying excitations. Employing Huygens principle, if the source (s) is located within a closed surface (S) such that (n) is the outward-normal to it, (Q) is a point on the surface, and (M) is a point in the field exterior to (S), then the desired solution of the nonhomogeneous wave equation (J-16) in terms of the Green's function is of the form

$$\Psi(M;t) = \int_S K(M,Q)\ \Phi(Q;t)\ d(Q)\ . \qquad \text{(J-19)}$$

$K(M,Q)$ is the Green's function and satisfies the differential equation

$$L\left[K(M,Q)\right] = 0 \qquad \text{(J-20)}$$

everywhere except at the point $M = Q$. Some of its other properties are that for fixed $Q, K(M,Q)$ is a continuous function of (M) and satisfies the prescribed boundary conditions. Except at the point $M = Q$, the first and second order derivatives of $K(M,Q)$ with respect to (M) are continuous. At the point $M = Q$, the first derivative has a finite discontinuity. The final form of Eq. (J-19) will depend on the prescribed boundary conditions and the type of excitation. This will define Green's function.

It is convenient to assume the source strength separable temporally, as well as spatially, that is,

$$\Phi(Q;t) = \Phi(Q)\ f(t)\ . \qquad \text{(J-21)}$$

$\Phi(Q)$ corresponds to the amplitude distribution of the source strength across the surface (S) and $f(t)$ is the applied time excitation. From the previous section, if $f(t)$ is harmonic, then the Fourier transform of $\Phi(Q;t)$ with respect to (Q) is the steady-state far-field radiation pattern. Whether or not measured sound fields actually represent steady-state values is particularly important when $f(t)$ is nonsinusoidal such as a modulated carrier. For the sound field surrounding a transducer to build up to its steady-state value, a sufficient time must be allowed for the sound waves to pass from one part of the transducer to another. The time required is of the order of the linear dimensions of the transducer divided by the velocity of sound. Note that the contribution to the observing point (M) at a time (t) from a source at a distance (r) is determined by the behavior of the source at an earlier time $(t - r/c)$. Thus, the time (T_{ss}) to establish the steady-state pattern for an aperture width (a) is approximately

$$T_{ss} \approx \frac{a + r}{c}\ . \qquad \text{(J-22)}$$

In section E, it was shown that there is a reciprocal relationship, similar to that in the time-frequency domain, between the width of the aperture distribution (a) and the half-power beamwidth (θ) of the corresponding radiation pattern. If (T_o) is the period of the carrier frequency of the excitation, Eq. (J-22) may be rewritten as

$$T_{ss} \approx \frac{T_o}{\theta} + \frac{r}{c}\ . \qquad \text{(J-23)}$$

For a superdirective array, the effective aperture width will encompass the physical dimensions of the array and the region which controls the extension of the pattern in the "inaccessible region" of imaginary angles.

If the field point (M) is a great distance from the source, then the pressure at that point is given by

$$p(M;t) = -\rho \frac{\partial\Psi(M;t)}{\partial t} \tag{J-24}$$

where (ρ) is the density of the medium. Substituting Eq. (J-19) into Eq. (J-24), where $\partial/\partial t \equiv$ partial time derivative

$$p(M;t) = -\rho \int_S K(M,Q) \frac{\partial}{\partial t}\left[\Phi(Q;t)\right] d(Q) \tag{J-25}$$

which is a generalized expression for the sound pressure at any point and instant of time in the acoustic far-field, and for an array having an arbitrary spatial configuration and temporal excitation.

HARMONIC EXCITATION

An important solution of Eq. (J-19) and Eq. (J-25) is when the excitation is harmonic with time. However, the ease with which the velocity potential or pressure is determined depends upon how the Green's function is defined. If $K(M,Q)$ satisfies the homogeneous boundary condition

$$\frac{\partial K(M,Q)}{\partial n} = 0 \tag{J-26}$$

on the closed surface (S), then in the far-field, $K(M,Q)$ has the form

$$f\left(\frac{M-Q}{|M-Q|}\right)\frac{e^{-j\frac{2\pi}{\lambda}|M-Q|}}{|M-Q|} \tag{J-27}$$

where $f(M,Q)$ is a function determined by the source configuration. For a piston-type transducer in an infinite rigid baffle,

$$K(M,Q) = \frac{e^{-j\frac{2\pi}{\lambda}|M-Q|}}{|M-Q|} \tag{J-28}$$

Since the time variation of the source strength corresponds to the single angular frequency (ω), the velocity potential Eq. (J-19) may then be written as

$$\Psi(M;t) = \int_S \frac{\Phi(Q)}{r} e^{j\omega(t-r/c)} d(Q) \tag{J-29}$$

where $|M-Q| = r$ and $\Phi(Q;t) = \Phi(Q)e^{j\omega t}$. Applying Eq. (J-24) to Eq. (J-29), the far-field pressure $p(M,t)$ becomes

$$p(M;t) = -j\omega\rho \int_S \frac{\Phi(Q)}{r} e^{j\omega(t-r/c)} d(Q)\ . \tag{J-30}$$

This reduces to Eq. (H-8) for a line source.

There are other methods for determining the far-field radiation pattern using the Green's function. With a periodic unit source at (Q), the pressure at the field point (M) may also be expressed as

$$p(M) = \frac{1}{4\pi} \int_S \left[p(Q) \frac{\partial}{\partial n} K(M,Q) - K(M,Q) \frac{\partial}{\partial n} p(Q) \right] dA(Q) \tag{J-31}$$

where $K(M,Q)$ is a solution of the inhomogeneous equation

$$\left[\nabla^2 + \left(\frac{2\pi}{\lambda} \right)^2 \right] K(M,Q) = -4\pi\delta(M,Q) . \tag{J-32}$$

In problems involving harmonic waves, the unit rotating vector $\exp(j\omega t)$ is often omitted, as in Eq. (J-31), but is nevertheless inferred. If $K(M,Q)$ is chosen as $\exp(j2\pi r/\lambda)/r$, then Eq. (J-31) becomes

$$p(M) = \frac{1}{4\pi} \int_S \left[p(Q) \frac{\partial}{\partial n} \left(\frac{e^{j\frac{2\pi}{\lambda} r}}{r} \right) - \frac{e^{j\frac{2\pi}{\lambda} r}}{r} \frac{\partial}{\partial n} p(Q) \right] dA(Q) \tag{J-33}$$

An approximation commonly made in the far-field is

$$\frac{\partial}{\partial n} \left(\frac{e^{j\frac{2\pi}{\lambda} r}}{r} \right) \approx j\frac{2\pi}{\lambda} \frac{e^{j\frac{2\pi}{\lambda} r}}{r} \frac{\partial r}{\partial n} = -j\frac{2\pi}{\lambda} \cos\theta \frac{e^{j\frac{2\pi}{\lambda} r}}{r} \tag{J-34}$$

where (θ) is the angle between the normal vector (n) to the surface (S) and the unit vector pointing from (Q) towards (M). Equation (J-33) may now be written as

$$p(M) = -\frac{1}{4\pi} \int_S \left[j\frac{2\pi}{\lambda} \cos\theta \, p(Q) + \frac{\partial p(Q)}{\partial n} \right] \frac{e^{j\frac{2\pi}{\lambda} r}}{r} dA(Q) \tag{J-35}$$

Though Eq. (J-35) is easy to evaluate, it requires knowledge of both the pressure $p(Q)$ and the normal gradient of the pressure $\partial p(Q)/\partial n$ over (S). However, if the Green's function is defined as the solution to Eq. (J-32) that vanishes over the surface (S), then Eq. (J-31) reduces to

$$p(M) = \frac{1}{4\pi} \int_S p(Q) \frac{\partial K(M,Q)}{\partial n} dA(Q) . \tag{J-36}$$

Equation (J-36) permits determining the pressure in the far-field, $p(M)$, in terms of measurements of the pressure over the source, $p(Q)$. Although it does not require a knowledge of $\partial p(Q)/\partial n$, $p(Q)$ and $\partial p(Q)/\partial n$ cannot be specified independently of one another over the surface. In all of these methods, both the amplitude and phase of the pressure over (S) must be determined. It is often convenient to measure the phase relative to the source excitation. If the radius of curvature of the transducer is equal to or less than a wavelength, and if the phase is slowly-varying across the surface, then as a first approximation the wave field in the immediate vicinity of the transducer may be considered as a plane wave propagating normally to the transducer face. As a result,

$$\frac{\partial p(Q)}{\partial n} \approx j\frac{2\pi}{\lambda} p(Q) \tag{J-37}$$

making Eq. (J-35) more attractive for determining $p(M)$. Other methods not involving Green's functions can be used employing other coordinate systems (Ref. J-13).

RESPONSE TO A MODULATED SIGNAL

An important problem in acoustics is the transmission and reception of wide-band signals. Of particular interest is the ability to characterize the information in a modulated signal. A modulated signal is one having some characteristic of it changing in a systematic manner which may or may not be enhanced by the spatial properties of the medium and array. For a linear array of discrete point elements, there may be a degradation of the radiation pattern if the modulation wavelength becomes comparable with the element spacing. If the modulation wavelength is less than the total length of the array, then at a specific instant of time, different elements of the array may transmit or respond to different portions of the modulated cycle.

Spatial properties may be influenced by time-modulating the electrical signal or some parameter of the spatial element, such as its length, excitation function, or number of elements. If the modulation is periodic, and the fundamental modulation frequency (ω_o) is much smaller than the carrier frequency (ω_c), then the time-varying radiation pattern may be written as

$$G(\theta; t) = A\left\{b_o(\theta) + b_1(\theta) \cos \omega_o t + \cdots\right\} e^{j\omega_c t} \tag{J-38}$$

where the $b_n(\theta)$ are patterns which depend only on the structural spatial properties and (A) is a constant. All harmonics $(n\omega_o)$ are independent of each other and are associated with specific radiation patterns $b_n(\theta)$. For example, if it is desired to form $2N+1$ pencil beams from a linear array of length (ℓ) where the spacing between beams is (θ_o), then Eq. (J-38) becomes

$$G(\theta; t) = \sum_{n=-N}^{N} \frac{\sin \frac{\pi}{\lambda} \ell(\nu - n\nu_o)}{(\nu - n\nu_o)} e^{j(\omega_c + n\omega_o)t} \tag{J-39}$$

where (ν) and (ν_o) are $\sin\theta$ and $\sin\theta_o$, respectively. A target in the vicinity of the angular direction $(n\nu_o)$ is directly associated with the frequency $(n\omega_o)$. Thus, Eq. (J-39) may be considered as representing a frequency spectrum in which the magnitudes of the upper and lower sidebands indicates the presence and magnitudes of the targets in the corresponding directions. These pattern characteristics are shown in Figure (J-1). By applying the Fourier transform to Eq. (J-39), the aperture distribution or excitation function is

$$F(x; t) = \sum_{n=-N}^{N} e^{-j\left(\frac{2\pi}{\lambda} n\nu_o x - n\omega_o t\right)} \tag{J-40}$$

which represents a train of standing waves traveling in the positive direction. Note that Eq. (J-40) becomes a delta function in the limit as N becomes infinite.

It has been shown that when the modulation bandwidth is small compared to the carrier frequency, then single-frequency analysis is usually sufficient. For other cases, it may be necessary to sum the signals from the various elements of the array as a function of time in the desired direction to obtain the net pressure. Another criterion, discussed by Pritchard in Ref. J-20, is the mean-square response which is particularly useful if the modulated signal contains a random component. The mean-square response is also a natural means of evaluating the directivity factor which is a spatial average of the sound intensity in transmission or of the square of the open-circuit output voltage from the array in reception.

When a signal is passed through an array, it will have the frequency and spatial characteristics of the array impressed upon its own frequency characteristics. If $F(j\omega)$ is the Fourier spectrum of the modulated signal, and $G(\theta,\omega)$ is the steady-state radiation pattern in the far-field, then the mean-square response of the array at any angle (θ) is proportional to

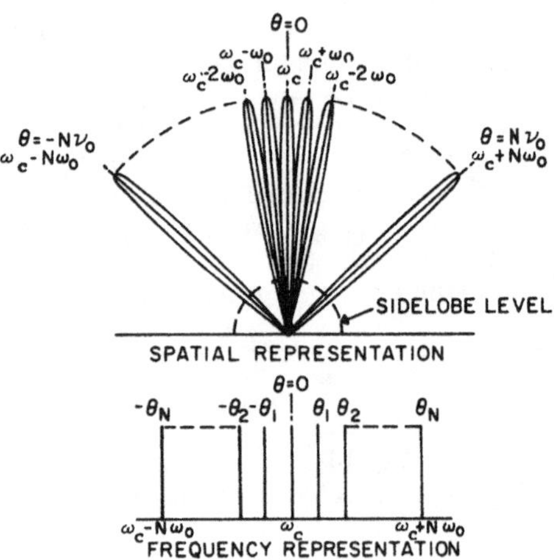

Figure J-1 - Time modulated linear array forming 2N+1 pencil beams, each of which is associated with a different frequency component

$$\int_{\omega_1}^{\omega_2} |F(j\omega)|^2 \, |G(\theta,\omega)|^2 \, d\omega \qquad \text{(J-41)}$$

where (ω_1) and (ω_2) are the lower and upper cut-off frequencies of the system, respectively. At $\theta = 0$, $G(\theta,\omega) = 1$, and Eq. (J-41) becomes

$$\int_{\omega_1}^{\omega_2} |F(j\omega)|^2 \, d\omega \qquad \text{(J-42)}$$

which from Plancherel's theorem, is the total energy in the modulated signal. The ratio of these two expressions is defined as the normalized mean-square response at the angle (θ), $\overline{G^2(\theta)}$,

$$\overline{G^2(\theta)} = \frac{1}{(\omega_2 - \omega_1)\,\overline{F(j\omega)^2}} \int_{\omega_1}^{\omega_2} |F(j\omega)|^2 \, |G(\theta,\omega)|^2 \, d\omega \qquad \text{(J-43)}$$

where

$$\overline{F(j\omega)^2} = \frac{1}{(\omega_2 - \omega_1)} \int_{\omega_1}^{\omega_2} |F(j\omega)|^2 \, d\omega \, . \qquad \text{(J-44)}$$

The m-s response removes all phase information and consequently is more suitable to amplitude modulation. However, a correlation in phase introduced by the modulation results in a time variation of the average energy and thus can be described by the m-s response.

For a given array and modulated signal, if the element spacing is less than the modulation wavelength, then the directional properties of the array will not be appreciably reduced. An order of magnitude for the modulation wavelength (λ_m) is the product of the velocity of sound in the medium and the period of the modulation. This may be approximated by

$$\lambda_m \approx \frac{4\pi c}{(\omega_2 - \omega_1)} \qquad \text{(J-45)}$$

If these conditions are met, then the level of the normalized m-s response for a modulated signal will be nearly that for a signal of a single frequency within the band.

When the modulation wavelength is much less than the element spacing, it is convenient to determine the response as a function of time and of angle. For a linear point source array at broadside, the response as a function of time will consist of a single signal having the same shape, but greater amplitude, than that emitted from or received by each element. As (θ) is increased, the beamwidth will increase and cause the response to be spread over a longer period of time. Due to interference between the signals from the separate elements, the amplitude response will be reduced relative to that at broadside. A further increase in (θ) will minimize the interference resulting in a time response comprised of a series of signals corresponding to each element. When this occurs, increasing (θ) will increase the distance between individual signals and consequently spread the response over a longer period of time. Furthermore, all amplitude levels of the response will remain fixed.

The preceding discussion excludes superdirective arrays. A superdirective array is more frequency sensitive and thus will lead to somewhat different results for the response to a modulated signal. If the array is designed for the midband frequency, then a narrow beam will be maintained but the level of the normalized mean-square response may be large for values of (θ) near $\pm\,\pi/2$. However, as before, if $F(j\omega)$ is symmetrical with respect to the midband frequency, then the major lobe of the m-s response for the modulated signal will be at least as narrow as the major lobe of the m-s response for a signal of a single frequency at the midband frequency.

4. SPACE-TIME CORRELATION FUNCTION

The ability of an array to discriminate against plane waves having spatially random phase characteristics is determined by the coherence of the waves at various points on the array. If the state of knowledge is the value of the pressure at two arbitrary points (X_1) and (X_2) and time instants (t_1) and (t_2), then the coherence may be described by averaging the product of the pressures either temporally, statistically, or spatially.

In the case of time averaging, the representation is called the space-time correlation function, $\psi(X_1, t_1; X_2, t_2)$ and is expressed as

$$\psi(X_1, t_1; X_2, t_2) = \langle p(X_1, t_1)\, p(X_2, t_2)\rangle \qquad \text{(J-46)}$$

where the angular bracket refers to a time average,

$$\langle \ldots \rangle = \lim_{T \to \infty} \frac{1}{2T} \int_{-T}^{T} \ldots \, dt \, . \qquad \text{(J-47)}$$

Statistical averaging corresponds to the different possible states of the medium. If the pressure wave is ergodic, time averaging and statistical averaging yield identical results, depending only on the time difference $\tau = t_2 - t_1$. For a spatially homogeneous process, the space-time correlation function depends only on the coordinate differences $x = X_2 - X_1$, $y = Y_2 - Y_1$, $z = Z_2 - Z_1$. Thus, for a stationary and spatially homogeneous random process, Eq. (J-46) becomes

$$\psi(X_1, t_1; X_2, t_2) = \psi(X_2 - X_1; \tau) = \langle p(X_1, t)\, p(X_2, t + \tau)\rangle \qquad \text{(J-48)}$$

and would yield equivalent results if averaged spatially. As the distance between points and the difference between time instants increases, the correlation function decreases.

If the acoustic pressure $p(x, y, z, t) = p(X, t)$ satisfies the wave equation

$$\nabla^2 p - \frac{1}{c^2} \frac{\partial^2 p}{\partial t^2} = 0 \, , \qquad \text{(J-49)}$$

then it follows that

$$\nabla_1^2 \psi - \frac{1}{c^2} \frac{\partial^2 \psi}{\partial t^2} = 0 \, ; \qquad \nabla_2^2 \psi - \frac{1}{c^2} \frac{\partial^2 \psi}{\partial t^2} = 0 \qquad \text{(J-50)}$$

where (∇_1) and (∇_2) are the vector derivatives with respect to (X_1) and (X_2), respectively. From Eq. (J-46), assuming stationarity, it follows that

$$\left[\psi(X_1, X_2, \tau)\right]^2 \leq \psi(X_1, X_1, 0)\, \psi(X_2, X_2, 0) \, , \qquad \text{(J-51)}$$

and by replacing (t) by $(t - \tau)$,

$$\psi(X_2, X_1, -\tau) = \psi(X_1, X_2, \tau) \qquad \text{(J-52)}$$

Equations (J-50), (J-51), and (J-52), together with appropriate boundary conditions, are sufficient for determining (ψ).

Sometimes it is more convenient to consider the Fourier transform of the correlation function,

$$W(X_1, X_2, \omega) = \int_{-\infty}^{\infty} \psi(X_1, X_2, \tau)\, e^{-j\omega\tau}\, d\tau \qquad \text{(J-53)}$$

which satisfies the equation

$$W(X_1, X_2, \omega) = W(X_2, X_1, -\omega)\,. \qquad \text{(J-54)}$$

The space-time correlation function and its transform are primarily useful for describing incoherent acoustic pressures and the effect of linear operations on them. For example, if the pressures at any two distinct points (X_1) and (X_2) are uncorrelated, then Eq. (J-48) reduces to

$$\psi(X_1, X_2, \tau) = \sigma^2(X_1)\, \delta(X_2 - X_1)\, \delta(\tau) \qquad \text{(J-55)}$$

where $\sigma^2(X_1)$ is the mean square amplitude at the point (X_1) and (δ) denotes the unit impulse function. Using the frequency description, it follows from Eq. (J-53) that

$$W(X_1, X_2, \omega) = \sigma^2(X_1)\, \delta(X_2 - X_1)\,. \qquad \text{(J-56)}$$

In general, if $r(X, t)\, dV$ is the impulse response of an array of linear spatial elements in volume dV, then the average power σ_o^2 at the output of the array may be written as

$$\sigma_o^2 = \iiiint r(X_1, t_1)\, r(X_2, t_2)\, \psi(X_1, X_2, t_2 - t_1)\, dV_1 dV_2\, dt_2 dt_1\,. \qquad \text{(J-57)}$$

By applying the convolution theorem to Eq. (J-57), the expression for the average output power becomes

$$\sigma_o^2 = \frac{1}{2\pi} \iiint H(X_1, \omega)\, H(X_2, \omega)\, W(X_2, X_1, \omega)\, dV_1\, dV_2\, d\omega \qquad \text{(J-58)}$$

where $H(X, \omega)$ is the transfer function of the array. All volume integrations are over the space (V) of the receiver elements. Optimizing the receiving response implies controlling the space-time correlation function as well as the impulse response. This may be done by selecting the configuration of the array, the manner in which the outputs are connected, and the frequency response of the elements.

As an example, consider an isotropic noise field, defined as one in which the total noise power received by a directional receiver is independent of both its location and its angular orientation. If the distance between two arbitrary points (X_1) and (X_2) is denoted by (r), a particular solution of Eq. (J-46) for the assumed conditions is

$$\psi(r, \tau) = \frac{c}{2r} \int_{\tau - r/c}^{\tau + r/c} \psi_o(\tau')\, d\tau'. \qquad \text{(J-59)}$$

$\psi(r, \tau)$ is the space-time correlation function of a uniform, isotropic noise field. From Eqs. (J-51) and (J-52), it is required that

$$\psi_o(\tau) = \psi_o(-\tau) \qquad \text{(J-60)}$$

and

$$\psi_o(\tau) \le \psi_o(0)\,, \qquad \text{(J-61)}$$

respectively. The function $\psi_o(\tau)$ is the autocorrelation function of the pressure at any point in the field,

$$\psi_o(\tau) = \psi(0,\tau) = \langle p(X,t)\,p(X,t+\tau)\rangle . \qquad \text{(J-62)}$$

Since this is related to the power spectrum, $W(\omega)$, of $p(X,t)$ by the Fourier cosine transform (Wiener-Khintchine theorem), Eq. (J-59) may also be expressed as

$$\psi(r,\tau) = \frac{1}{2\pi}\int_0^\infty W(\omega)\,\frac{\sin\frac{\omega r}{c}}{\frac{\omega r}{c}}\cos\omega\tau\,d\omega \qquad \text{(J-63)}$$

where $W(\omega)$ is positive for all (ω). If $\psi_o(\tau)$ or $W(\omega)$ can be expressed analytically, $\psi(r,\tau)$ may be determined from either Eq. (J-59) or Eq. (J-63) using contour integration.

For the special case of a rectangular noise spectrum one octave in width, and a linear array parallel to the x-axis, Eq. (J-63) becomes

$$\psi(x,0) = \frac{S_i(4\pi x) - S_i(2\pi x)}{2\pi x} \qquad \text{(J-64)}$$

where (x) is the spacing in wavelengths and $S_i(x)$ denotes the sine-integral. Figure (J-2) shows a plot of $\psi(x,0)$ as a function of the separation in wavelengths at the lower cutoff frequency. Note that it is essentially zero when (x) is greater than 0.348. It may be generalized that the effective directivity factor for uniform noise for a set of N receivers arranged arbitrarily in space is $1/N$, provided that the spacing between adjacent elements is equal to or greater than a half-wavelength at the geometric mean frequency of the band.

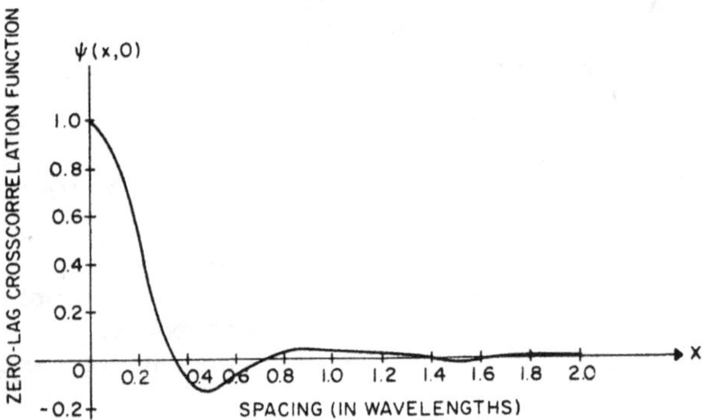

Figure J-2 - Crosscorrelation of octave band noise

5. DIRECTIVITY FACTOR

The utility of the directivity factor is that it is a measure of the ability of a directional element to discriminate against interference arriving from bearings other than those desired since its response to any distributed interference will be less than the response of a nondirectional element. The directivity factor will depend on the signal-to-noise ratio generated at the output of the element and is a function of the geometrical configuration of the array, frequency of the received signal, and the distribution of noise sources in the medium.

The manner in which the transmitted acoustic energy at a fixed bearing varies with frequency or the manner in which the energy distribution generated by hydrophones varies with frequency when receiving acoustic waves over any given bearing, depends on the frequency dependence of the radiation pattern. A measure of this is the manner in which the directivity factor (D.F.) varies with frequency. For the detection of signals of the threshold condition, enhancing the directivity factor by the process of "shading" or selecting the element sensitivities may be carried out with regard only to the background noise. For such applications array design is not dependent upon the spectrum of the signal, as it is for a transmitting array, but only on the spectrum of the noise at the output of the array. If the receiver does not contain any nonlinear elements, then the passband need only be as wide as is necessary to handle the signal.

In section H-7, it was indicated that by proper selection of the excitation coefficients and element spacing, a directivity factor greater than that for a uniform, in-phase excitation may be obtained. If the spacing is sufficiently wide so that the correlations of the noise received by the elements is minimized, then having sensitivities other than uniform will do little to increase the directivity factor. For element spacings less than a half-wavelength, a significant increase in the directivity factor can be effected, but the increase is less at wide bandwidths. In general, the maximum directivity factor of an equally-spaced broadside array for wide-band operation is less than that for single-frequency operation, much more than that obtained by use of the single-frequency design at the wide bandwidth, and except for very wide-band cases, is significantly greater than that obtained by using uniform excitation. When a wide-band source is moved off the the maximum response axis (MRA) of an array, the effect is to broaden the pattern function. Since the directivity factor increases with increase in frequency, resulting in a narrower beam, the effect of having a wide-band signal off the MRA is to enhance the low-frequency components with respect to the high-frequency components.

For a linear array, the relationship between the directivity factor and pattern function $G_{2N+1}(\theta)$ is

$$\frac{1}{\text{D.F.}} = \frac{1}{2} \int_{-\pi/2}^{\pi/2} \left[G_{2N+1}(\theta)\right]^2 \cos\theta \, d\theta \, . \tag{J-65}$$

If the received signal is modulated, then the squared bracketed term refers to the mean-square response defined by Eq. (J-43). Substituting this into Eq. (J-65), the directivity factor for modulated signals $\overline{(\text{D.F.})}$ is related to the directivity factor for single-frequency operation (D.F.) by

$$\overline{(1/\text{D.F.})} = \frac{1}{(\omega_2 - \omega_1)\,\overline{F^2(\omega)}} \int_{\omega_1}^{\omega_2} |F(\omega)|^2 \frac{1}{\text{D.F.}(\omega)} \, d\omega \, . \tag{J-66}$$

In the special case of white noise, $|F(\omega)|^2$ is a constant independent of frequency, and Eq. (J-66) reduces to

$$\overline{(1/\text{D.F.})} = \frac{1}{(\omega_2 - \omega_1)} \int_{\omega_1}^{\omega_2} \frac{1}{\text{D.F.}(\omega)} \, d\omega \, . \tag{J-67}$$

If the directivity factor is proportional to all frequencies contained in the noise spectrum, then it is but slightly reduced relative to the directivity factor for a signal at a single frequency at the center of the band, $\omega_o = (\omega_1 + \omega_2)/2$. In most cases, if the modulation frequency of a single target on the MRA is much lower than the center frequency of the band, there will be no appreciable change of the directivity factor regardless of the target spectrum. For multiple targets, classification of the modulation will depend upon how different or orthogonal the target spectra are, relative to each other, as well as upon the modulation frequency.

For a directional element in a uniform isotropic noise field, the directivity factor (D.F.) may also be defined as the value of the signal-to-noise ratio characteristic of the element's location, $(S/N)_{ref}$, divided by the actual value of the signal-to-noise ratio, $(S/N)_{act}$, at the output of the element. Thus,

$$\text{D.F.} = \frac{\left(\frac{S}{N}\right)_{ref}}{\left(\frac{S}{N}\right)_{act}} \tag{J-68}$$

where each factor in the ratio represents a power per unit band at some specified signal frequency. The signal-to-noise ratio characteristic is that pertaining to a hypothetical nondirectional element whose receiving response is equal to the maximum receiving response of the given directional element. If the directional element is receiving a signal on its MRA, then the responses of both elements to this signal will be the same. Equation (J-68), expressed in decibels, then becomes

$$N_{act}(\text{db}) = N_{ref}(\text{db}) - \text{D.I.} \tag{J-69}$$

where D.I. is the directivity index defined in Section (H-7).

Equation (J-68) assumes that the acoustic interference is so distributed that the power per unit band is the same at all bearings. However, in practice, the distribution is not uniform. This is taken into account by defining an effective directivity factor (EDF) as the ratio of the power per unit band available from a directional element to the power per unit band available from a nondirectional element at the same location. As a result of the directionality of the interference, the EDF in any given location is quite likely to be a function of the orientation of the element.

Because it is easier to determine the response of a spatial array to plane waves, if the acoustic waves actually received are not plane, then they are represented by equivalent plane waves, with the equivalence being on the basis of energy—phase effects are not considered. For any specified frequency, the apparent plane wave intensity per unit band of the acoustic energy received by a directional element is defined as the acoustic energy of plane waves which if received on the MRA of the element, would generate the same power per unit band in the output as the actual waves. The intensity per unit band is the third derivative of energy with respect to time, bandwidth, and area. For a nondirectional element, the apparent plane wave intensity per unit band is called the equivalent plane wave intensity per unit band. If the directional and nondirectional elements have equal area, then the directivity factor may be expressed as

$$\text{D.F.} = \frac{\text{A.P.W.I.}}{\text{E.P.W.I.}} \tag{J-70}$$

For a directional noise field, the ratio in Eq. (J-70) would equal the effective directivity factor.

6. SPACE-TIME SAMPLING

The concept of sampling has been discussed separately for the time and space domains. A natural extension of theory seems to be that of sampling jointly in space and time, and determining the effect of boundary and initial conditions upon the total information content. A useful approach from the point of visualization is to examine the acoustic wave equation.

The acoustic wave equation is a linear partial differential equation which describes continuous flow in the medium. If the acoustic process and medium can be considered a continuous system, normal mode theory shows that an arbitrary vibration may be represented as a superposition of eigenvibrations, that is, vibrations for which the dependent variable is expressed as the product of a factor depending only on the time, and a factor depending only on the spatial position. If operated linearly upon, the eigenvibration will yield the same vibration multiplied by a constant, the eigenvalue. The associated eigenvalues form an infinite sequence of space-time samples which completely characterize the medium. By imposing boundary and initial conditions, the wave equation may be effectively replaced by a difference equation having a finite sequence of eigenvalues or space-time values. This implies that the spatial and temporal constraints have transformed a continuous field to a system having a finite number of degrees of freedom.

If an array is limited in time operation and frequency response to (T) and (W), respectively, the received pressure distribution at any point on the aperture can be represented by $2WT$ sampled values. Similarly, if (θ) is the half-power beamwidth and (L) the length of the aperture parallel to some axis, the one-dimensional distribution at any instant of time can be represented by $\theta L/\lambda$ sampled values across the aperture. The factor of two is unnecessary here because knowledge about phase at sample points spaced λ/θ apart on the aperture is obtained from the $2WT$ sample points in time. Assuming the field to be comprised of eigenvibrations, the total number of sampled values may be taken as the product of the spatial and temporal degrees of freedom. Thus, the received distribution across the aperture (L) during time (T) may be represented by $N_s = 2WT\theta L/\lambda$ sampled values. However, it was indicated that a sufficient time must be allowed for the sound waves to pass from one part of the transducer to another. The time required is of the order of the linear dimensions of the transducer divided by the velocity of sound. Increasing the transducer dimensions to increase angular resolution constrains the minimum duration signal which the transducer can accommodate.

If (H) possible amplitudes of the pressure can be measured at each sample point, then for the one-dimensional aperture, the total number of different signals which can be described by (N_s) sample values is

$$H^{(2WT\theta L/\lambda)} \tag{J-71}$$

Generalized, if there are H_q specifiable values of a physical parameter at each sample point, then for a three-dimensional array of volume (V), having "beamwidths" θ_x, θ_y and θ_z along the cartesian axes, there are

$$\prod_{i=1}^{q} H_i^{(2WTV\theta_x\theta_y\theta_z/\lambda^3)} \tag{J-72}$$

possible signals. Any one of these signals can give the history throughout the volume (V) and for the time (T), but there is only one which gives the actual history of the space-time domain. Equation (J-72) gives the maximum permissible signals to be described by $N_s = (2WTV\theta_x\theta_y\theta_z/\lambda^3)$ sample values. The actual number of signals which are observable is ordinarily much less due to constraints which reduce the effective number of independent dimensions and the range of observable or specifiable values.

7. RESOLUTION

The type of resolution discussed here is that of angular resolution, also referred to as the resolving power, which may be determined from the radiation pattern of a receiving element. The complete details of the radiation pattern may be affected by spatial and temporal parameters, both deterministic and statistical, and hence, the actual resolving power of a spatial element may not be simply defined. Since angular resolution is dependent on the signal-to-noise ratio of the received signal, the size and shape of the source and receiver, the distance between them, and the dynamic range of the receiving and display system, it cannot be described only in geometrical terms.

Resolving power is often expressed in terms of the beamwidth which is affected by the factors given above. For example, as a result of the inverse relation between the aperture width (in wavelengths) and beamwidth, resolution may be determined first by selecting the wavelength and size of aperture. Another important consideration is the level of the sidelobes. The presence of sidelobes affects the resolution of the system when the problem consists of the identification of two or more sources which have a range of intensities. Conventionally, equal intensities are assumed and the sidelobe levels are hence not taken into account. For a given aperture, sidelobes may be suppressed but at the expense of sacrificing beamwidth. Therefore, it is insufficient to merely state the beamwidth without including a description of the sidelobe structure.

The resolving power may also be significantly modified by nonlinearities in the system. The effects of nonlinearities will also depend on the dynamic range of the overall receiving and display system. In general, propagation variations which cause target scintillation, along with system instabilities modify the resolving power obtainable and establish the limit on how accurately it may be determined.

Information theory has influenced the classical concept of resolving power, first formulated in optics, by providing methods for determining the finite number of independent data contained in the response of a given spatial element, and comparing these with the information potentially available. Resolving power may thus be associated with space-time sampling.

Physically, the image of a point source is not created by the intersection of rays from the source, but results from the diffraction pattern of waves from the source that pass through the aperture plane. An optical system is said to be able to resolve two point sources when the corresponding diffraction patterns are sufficiently small or sufficiently separated to be distinguished. The numerical measure of the ability of the system to resolve two such points is called its resolving power. It increases with the solid angle of the cone of rays intercepted by the instrument and is inversely-proportional to the wavelength of the radiation used.

The criterion originally proposed by Lord Rayleigh is that two point sources are just resolvable if the central maximum of the diffraction pattern of one source just coincides with the first minimum of the other. This is illustrated in Figure (J-3) where (P_1) and (P_2) are two point objects, and (P_1') and (P_2') are the centers of their diffraction patterns, formed by some optical instrument represented in the diagram by a single lens. If the images are just resolved, then the separation of the centers of the patterns equals the radius of the central bright disk. From a knowledge of the focal lengths and separations of the lenses in any particular instrument, the distance between the two point objects may be computed. This distance is called the limit of resolution of the instrument. For the case of an instrument having a circular pupil of diameter (D), Rayleigh's criterion gives the minimum angular separation (R) of two points as

$$R = 1.22 \frac{\lambda}{D} \tag{J-73}$$

Note that this was obtained without considering the signal-to-noise ratio.

Since radiation from antennas and other spatial transducers is similar to the diffraction of light, Rayleigh's criterion has been used in determining their resolution capability. For a symmetrical radiation pattern, the Rayleigh resolution (R) is equal to one-half of the beamwidth between first nulls (BWFN), that is,

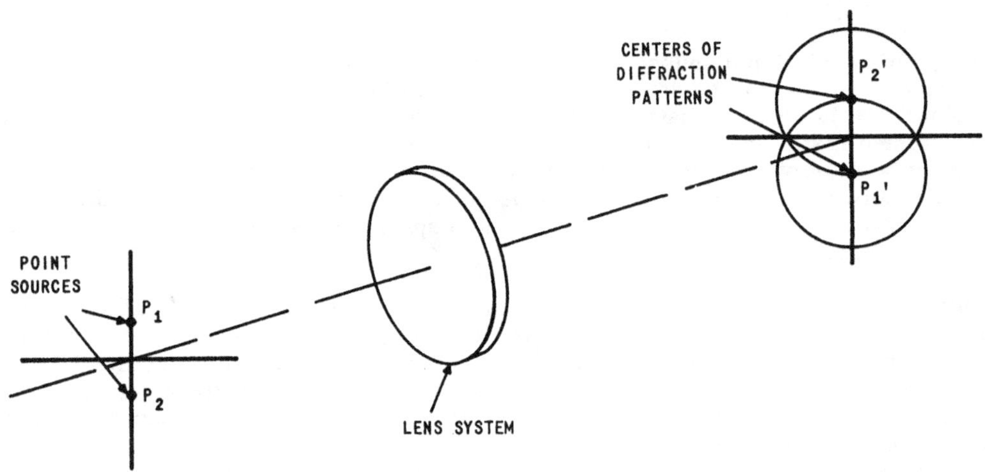

Figure J-3 - Illustration of the Rayleigh criterion

$$R = \frac{BWFN}{2} \tag{J-74}$$

A radiation pattern for a single point source is shown in Figure (J-4.a). The power pattern for two identical point sources separated by the Rayleigh angle is given by the solid curve in Figure (J-4.b), with the pattern for each source when observed individually shown by the dashed curves. It is seen that the two sources will be resolved provided the half-power beamwidth (HPBW) is less than one-half of the beamwidth between first nulls, which is usually the case. For a spatial element at broadside, having uniform excitation and of length (L),

$$BWFN = \frac{114.6}{L/\lambda} \tag{J-75}$$

and

$$HPBW = \frac{50.8}{L/\lambda} \tag{J-76}$$

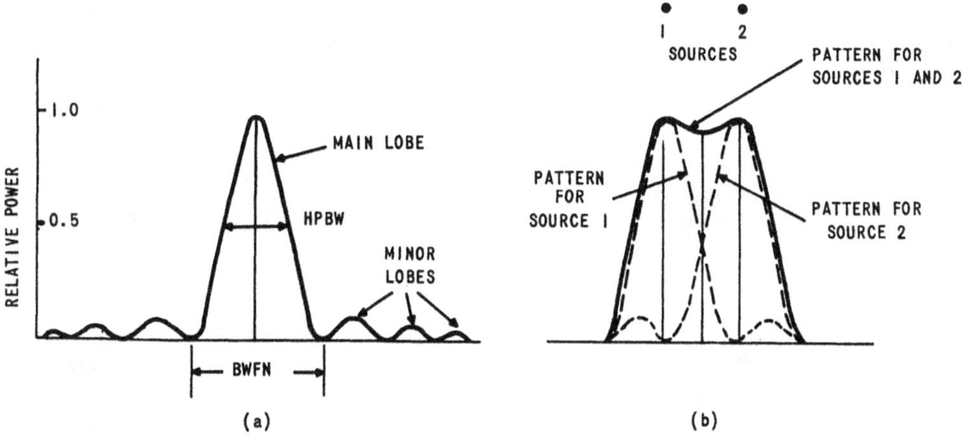

Figure J-4 - (a) Power pattern and (b) power patterns for two identical point sources separated by the Rayleigh angle as observed individually (dashed) and together (solid)

Using Eq. (J-74), the angular resolution (R) is given by

$$R = 1.12 \text{ HPBW} \tag{J-77}$$

where it is assumed that $L/\lambda >> 1$. Clearly, as (R) decreases, the resolving power increases.

The resolution capabilities of acoustic arrays are on the order of a beamwidth which is related to the size of the array aperture. A means of decreasing the beamwidth or improving the resolution is by using a superdirective array. This technique implies a very high Q, defined as 2π times the energy stored in the immediate region of the radiator divided by the energy radiated in one period, and will result in a reduced efficiency.

A method for designing and evaluating a superdirective array has been given by Tucker (Ref. J-37). The performance of the overall system is evaluated by a Noise Factor which is defined as the ratio of the signal-to-noise of an array with uniform in-phase sensitivity to the signal-to-noise of the excitation function needed to give a desired directivity index. This may be done by superimposing a number of elementary radiation patterns of the form $(\sin x/x)$. The number used will depend on the degree of accuracy desired. By applying a continuous phase shift to successive elements of the array, using electrical delay networks, a pattern can be shifted to either side of the axis normal to the array and will take the form

$$[\sin(x \pm n\pi)/(x \pm n\pi)] .$$

Therefore, as indicated in section (H-7), the desired pattern may be obtained by superimposing patterns whose zeros coincide and which have their peak responses in the region of imaginary angles. Their secondary lobes are then used to cancel out the secondary lobes of the ordinary $(\sin x/x)$ response. This is shown in Figure (J-5) for a linear array of length $\ell = 2\lambda$.

In Figure (J-5), the Noise Factor was found to increase in the synthesis of a superdirective response. In general, it is found that the improvement of directivity factor by the superposition of patterns with peaks in the range of imaginary angles adds to the noise output, while contributing nothing to the peak signal output. More important, the Noise Factor is worsened at a much faster rate than the directivity factor is improved.

Whether or not superdirective concepts can be used for receiving arrays strongly depends on the magnitude of background noise against which the signal is to be detected. The thermal-agitation noise of the dissipation resistance of the array is proportional to the power gathered by the whole radiation pattern including imaginary angles, whereas noise arising in the medium (thermal-agitation noise in water and "sea-state noise") is received by the portion of the pattern corresponding to real angles only. Consequently, if the limiting factor is the noise arising in the medium, then superdirectivity may give an improvement in the overall performance. However, if the system performance is dominated by the noise from the dissipation resistance

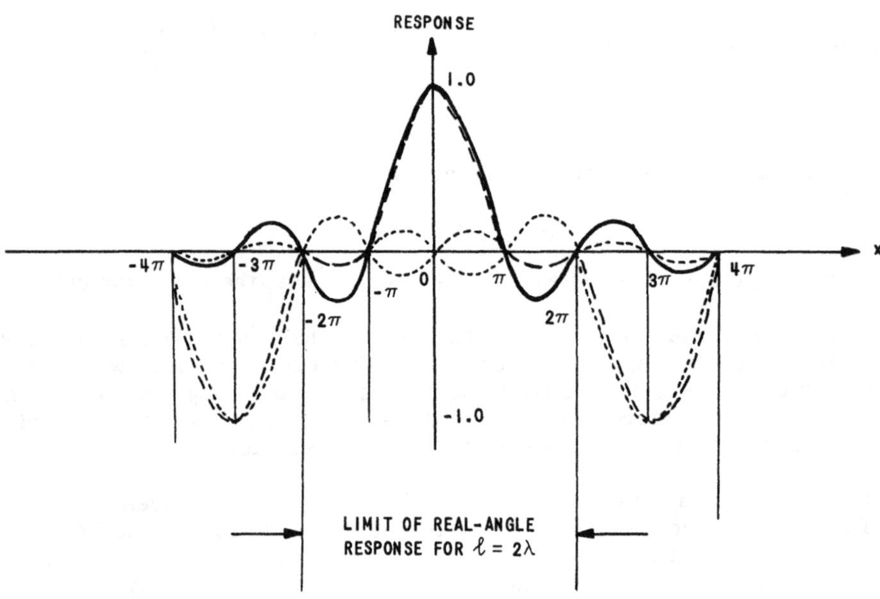

Figure J-5 - Synthesis of super-directive response with reduced secondary lobes and slightly narrowed main beams

of the transducer elements (increased by the interaction of the acoustic radiation impedance of the elements that make up the array) or from that in the receiving amplifiers, then the superdirective array is inferior to the ordinary array. The maximum value of the Noise Factor that can be obtained is that due to a uniform in-phase excitation.

A few examples have been given which illustrated the nature and interrelationship of the constraints. One of the effects of physical boundaries on acoustical processes was seen to transform a continuous "field" into a system having a finite number of degrees of freedom even when dissipative processes were not considered. In determining energy and intensity relationships of the system, the averaging time must be taken into account — and the details of the instrument should include its spatial extent.

Superdirectivity, and analogous concepts have been used to illustrate other relationships which may arise. It was indicated that a complete specification of superdirective configurations should include the accessible and the inaccessible regions. The net result is that although directional patterns having a narrower beam than may be obtained from uniform excitation of an aperture of the same width, the realization of such patterns is characterized by severe restrictions of bandwidth, and of susceptibility to inhomogeneities in the medium in the vicinity of the array.

All real processes are bounded and constrained in one or more domains, and important clues leading to improved understanding and classification of acoustic and electromagnetic space-time processes lie in the examination of the interrelationships and interactions which arise among the variables. Effective analysis will include these completely, thereby permitting the most economical utilization of the domains involved. Solutions to specific problems always necessitate detailed knowledge of the physical environment. However, useful results are derivable from the generalized concept that finite bounds exist on the total information, on the rate, and density. Physical processes may not be observed in infinite detail while still preserving the features upon which the identification or observation rules are based.

REFERENCES

J-1. P. G. Frank and A. Yaspan, "Wave Acoustics," chapter 2 of "Physics of Sound in the Sea, Part I: Transmission," reprinted and distributed by the National Research Council.

J-2. Lord Rayleigh, "Theory of Sound," Vol. 2, Macmillan Co., 1940.

J-3. "Basic Principles of Underwater Sound," Part I of "Principles of Underwater Sound," reprinted and distributed by the National Research Council.

J-4. L. E. Kinsler and A. R. Frey, "Fundamentals of Acoustics," Chapters 5 and 7, John Wiley and Sons, Inc., 1961.

J-5. C. B. Officer, "Introduction to the Theory of Sound Transmission," Chapters 1 and 2, McGraw-Hill Book Company, Inc., 1958.

J-6. V. H. Rumsey, "Some New Forms of Huygens' Principle," I.R.E. Trans. on Antennas and Propagation, December 1959, pp. S103-S116.

J-7. R. Hills, Jr., "Synthesis of Directivity Patterns of Acoustic Line Sources," Tech. Memo. 23, Acoustics Research Laboratory, Harvard University, November 1, 1951, Chapter VI.

J-8. P. C. Clemmow, "Infinite Integral Transforms in Diffraction Theory," I.R.E. Trans. on Antennas and Propagation, December 1959, pp. S7-S11.

J-9. Courant and Hilbert, "Methods of Mathematical Physics," Volume 1, Interscience Publishers, Inc., N.Y., 1955, pp. 275-286 and pp. 351-388.

J-10. L. Page, "Introduction to Theoretical Physics," D. Van Nostrand Company, Inc., Princeton, N.J., 1935.

J-11. C. W. Horton and G. S. Innis, Jr., "The Computation of Far-Field Radiation Patterns from Measurements Made Near the Source," J. Acoustical Soc. Am., Vol. 33, No. 7, July 1961, pp. 877-880.

J-12. L. L. Foldy and H. Primakoff, "Generalized Theory of Electroacoustic Transducers," Chapter 3 of "Basic Methods for the Calibration of Sonar Equipment," Tech. Rept. Division 6, NDRC, 1946.

J-13. J. Pachner, "Instantaneous Directivity Patterns," J. Acoust. Soc. Am., Vol. 28, 1956.

J-14. G. Chertock, "Sound Radiation from Prolate Spheroids," J. Acoust. Soc. Am., Vol. 33, No. 7, July 1961, pp. 871-876.

J-15. H. E. Shanks, "A New Technique for Electronic Scanning," I.R.E. Trans., Vol. AP-9, No. 2, March 1961, pp. 162-166.

J-16. H. E. Shanks and R. W. Bickmore, "Four-dimensional Electromagnetic Radiators," Canad. J. Physics, Vol. 37, March 1959, pp. 263-275.

J-17. C. Polk, "Transient Behavior of Aperture Antennas," I.R.E. Proc., Vol. 48, No. 7, July 1960, pp. 1281-1288.

J-18. H. J. Schmitt, "Transients in Cylindrical Antenna," Cruft Lab., Harvard University, Cambridge, Mass., Tech. Rept. No. 296, January 5, 1959.

J-19. G. G. Weill, "Transients on a Linear Antenna," California Institute of Technology, Pasadena, Antenna Lab., Tech. Rept. No. 20, August 1959.

J-20. R. L. Pritchard, "Directivity of Acoustic Linear Point Arrays," Tech. Memo. 21, Acoustics Research Laboratory, Harvard University, January 15, 1951, Chapter V.

J-21. C. H. Page, "Applications of the Fourier Integral in Physical Science," I.R.E. Trans., Vol. CT-2, No. 3, September 1955, pp. 231-237.

J-22. C. Eckart, "The Theory of Noise in Continuous Media," J. Acoust. Soc. Am., Vol. 25, No. 2, March 1953, pp. 195-199.

J-23. H. W. Marsh, Jr., "Correlation in Wave Fields," USL Quarterly Report, Part II, 1 October 1949 - 31 March 1950, pp. 63-68.

J-24. H. S. Heaps, "General Theory for the Synthesis of Hydrophone Arrays," J. Acoust. Soc. Am., Vol. 32, No. 3, March 1960, pp. 356-363.

J-25. J. J. Faran and R. Hills, "Wide-band Directivity of Receiving Arrays," Acoustics Research Laboratory, Harvard University, Tech. Memo. No. 31, May 1953.

J-26. J. W. Horton, "Fundamentals of Sonar," United States Naval Institute, 1957, pp. 217-221 and pp. 228-236.

J-27. R. Manasse, "An Analysis of Angular Accuracies From Radar Measurements," Astia No. 236166, 1960.

J-28. S. Goldman, "Information Theory," Prentice-Hall, Inc., New York, 1955, pp. 294-300.

J-29. L. L. Foldy, "Testing Technique," Chapter 5 of "Basic Methods for the Calibration of Sonar Equipment," Tech. Rept. Division 6, NDRC, 1946.

J-30. "Calibration of Electroacoustic Transducers," American Standards Association, 1958.

J-31. P. M. Morse, "Vibration and Sound," McGraw-Hill Book Company, Inc., New York, 1948.

J-32. W. Beranek, "Acoustic Measurements," John Wiley and Sons, Inc., New York, 1949.

J-33. W. M. Ewing, F. Press, and W. S. Jardetzky, "Elastic Waves in Layered Media," McGraw-Hill Book Company, Inc., New York, 1957.

J-34. B. B. Baker and E. T. Copson, "The Mathematical Theory of Huygens' Principle," Oxford University Press, New York, 1939.

J-35. C. Eckart, "Approximate Solutions of One-dimensional Wave Equations," Rev. Mod. Phys., Vol. 20, 1948, pp. 399-417.

J-36. J. Freedman, "Resolution in Radar Systems," I.R.E. Proc., July 1951, pp. 813-818.

J-37. D. G. Tucker, "Signal/Noise Performance of Super-Directive Arrays," Acustica, Vol. 8, 1958, pp. 112-116.

J-38. D. G. Tucker, "The Signal/Noise Performance of Electro-Acoustic Strip Arrays," Acustica, Vol. 8, 1958, pp. 53-62.

K. SUMMARY

A characteristic associated with the assembly, study, evaluation, and possibly the use, of the material received within this document, appears to be the tendency to oscillate between the extremes represented by the following quotations:

"The human understanding, from its peculiar nature, easily supposes a greater degree of order and equality in things than it really finds."

"Faced with too strong a flow of information, man can filter and recode up to a certain degree; soon he switches to random sampling of information; and from there to complete confusion is but one step."

At times there has been a strong temptation to "suppose a greater degree of order and equality" (and of utility) in the development of analogies and dichotomies, involving probabilistic and deterministic descriptions for acoustics and electromagnetics, and in the examination of physical processes as information, communications or decision channels. Equally strong at other times has been a sense of "complete confusion" associated with the attempts to keep pace with the many spoken and written pronouncements and the instrumental contributions which are steadily increasing year by year. Whatever "order and equality" exist, are largely and necessarily within generalized relationships rather than in specific details. Once this restriction is recognized and accepted, it is then possible to derive pragmatic guidelines to questions which should be raised and which must be answered when specific problems are analyzed. Examples of such questions are:

What are the basic commodities associated with the information processes? Have they been described completely and efficiently in terms which may be observed and measured? Are the mathematical concepts consistent with the physical processes? What are the possible consequences of uncertainties, exclusions, and constraints?

A brief review will be made of some of the major elements of the entire survey in order to illustrate the nature, use, and limitations of these guidelines.

In the initial sections a number of descriptions of functions were discussed. An important class involved statistical representations. Such descriptions may arise in problems where the only available data are statistical in nature, or in problems where it is convenient and possible to use averages, or probability distributions. For example, when large numbers of states exist, if all the details were to be included the resulting complexity of the representation may be too great to be useful. The basis for statistical descriptions rests ultimately on the validity of assumptions made of regularity and stationarity within relative scales of time and precision characterizing a particular problem. Probability distribution functions, and probability density functions — the derivative of the distribution — are commonly used. The characteristic function, which is the Fourier transform of the density function provides another measure. One of its features in conjunction with multiple variables is that by working with the characteristic function n-fold integrations may be replaced by multiplication operations. Practically, the distribution density, and characteristic functions provide more readily derivable and simplified measures such as the (averages) moments of the positive integral powers of the random variables.

Although readily derivable and useful in many operations, averages do not serve to "represent" signals, or processes. Statistical measures which retain utilitarian features and which do contain other features of signals and their interrelationships are correlation functions and power spectra. Correlation relates the linear dependence between two variables, and if the correlation is zero, the variables are linearly independent — however, a more detailed examination is required in order to determine whether the variables are statistically independent.

The examination may encompass higher moments of the probability density function. The mean spectral correlation function establishes the relation between phase coherence and the variation of the mean-square value of a variable as a function of time. The mean power spectrum is related to the correlation function through the Fourier transform, and the use of correlation or spectral analysis often may depend on instrumentation considerations, since in theory, the conjugate relationship indicates that they are "equivalent" descriptions. Thus, when working with very low audio frequencies, correlators which employ delay lines may be built more economically than spectrum analyzers.

As noted earlier, statistical and probabilistic descriptions are employed to simplify representations involving large numbers of states particularly when the descriptions can be made numerically. Because they are not strongly associated with specific features of signals and physical processes, analysis techniques may be and have been drawn from many areas. This does not minimize, however, the requirements for making a detailed physical analysis of each application to ensure, for example, that variables may be treated as being statistically independent. This requirement is even more important when other generalized measures such as entropy and information are employed. Entropy and informational concepts permits quantitative statements regarding probabilities of attributes regardless of what the attributes may be.

Although statistical relationships have been stressed, predictable or deterministic functions are often used — in fact, most problems involve an interplay among variables which require both types. Of great importance as tools for representation and analysis of both types are Fourier methods. The primary descriptions of a periodic function are its Fourier series and line spectra. The former is a decomposition into linearly additive trigonometric terms $e^{j\omega t}$ which indicate how the various harmonic terms contribute to the overall structure of the function. If the coefficients of a harmonic series expansion are chosen to be Fourier coefficients, then the mean square error will have its smallest possible value. Since this criterion results in Parseval's theorem — Fourier representations show how energy is distributed in frequency. Despite widespread use in analysis there is something disturbing about structural elements which are supposed to last forever. Real signals are, of course, limited in their duration, and real components are restricted in bandwidth. The imposition of an exact bandwidth limit implies that the only values of the function taken into account are those at certain times. Hence, different functions which take on the same values at these times are not distinguishable after being processed by an ideal filter. Although a bandwidth limitation permits a continuous function to be replaced by an enumerable sequence of sample value, ambiguities are "created" in the process. Similarly if the limit is in time, then a continuous spectrum may be expressed in terms of an equivalent line spectrum — and the only values of its spectrum which are taken into account are those at discrete frequencies.

For a signal which is restricted in time (T) and bandwidth (W), a fundamental principle states that $2WT$ sample values may be used to represent the signal. This characterization is only approximately true since there will be energy associated with the signal beyond the time interval (T) and the band (W). The magnitude of this energy is a measure of the error, and although its magnitude relative to the total energy may be small especially for large values of W and T, there still may be sizeable effects produced, for example, at the boundaries.

The introduction of a boundary in one domain quantizes the variable in the conjugate domain, and creates the possibility of ambiguities — that is, restricts the ability to make a unique identification, although it does permit replacing a continuous function with a set of values. Boundaries imposed jointly in conjugate domains, such as time and frequency, also restrict performing certain functions simultaneously. Thus, the ability to determine the arrival time and frequency content of a pulse which has been propagated by a band-limited channel is established by the exact nature of the boundaries, and in any event may be observed only to the extent permitted by the relationship $\Delta t \, \Delta f \approx 1$, where Δt and Δf are the indeterminacies in time and frequency.

It is not obvious that from these simple beginnings it would be possible to obtain much of practical value for the solution of problems of greater complexity characterized by higher dimensionality, multiple boundaries, and with a requirement for performing and assessing multiple tasks. Fortunately, some of the basic concepts appear to be extensible to such cases.

Early discussions indicated that the use of Fourier representations was appropriate for time-invariant processes since functions could be simply represented analytically and readily instrumented. Not only representations but operations were also simplified — descriptions of physical element behavior which involved the solutions of linear, differential equations were reduced to simple, algebraic problems. Fourier methods illustrate the basic philosophical elements of representation and transformation theory. These elements provide useful guidelines in many analysis and synthesis problems, and are extensible — holding equally well for circuit and spatial element analysis. The concepts include: simplification, matching, and invariance of structure under transformation. Some limitations were also recognized and although not included explicitly in the analysis, in transforming from one domain to another there are problems associated with the transformations of errors and of equipment tolerances. For spatial problems, Fourier methods are restricted to regions where the wavefront does not change with the independent spatial variable.

When the input to a constant coefficient linear component is a periodic function of time or space having a Fourier series which converges, the steady-state output will be another periodic function of time or space whose Fourier series is that of the input modified only in amplitude and phase. The effect of the component on the input is given by the transfer function. A linear component may also be described by its response to an impulse. The output will then be a pulse whose spectrum inherits the amplitude and phase characteristics of the network. For a spatial element, the analogy is the response of the element to a point source. The impulse response and transfer function (also called system function) are Fourier conjugates. When the input to a linear element is a transient, the spectrum of the output is the product of the spectrum of the input and the system function.

In applications the use of impulses (or point sources for spatial elements) is ordinarily more difficult to instrument than the instrumentation and the experimental procedures required to determine the system function. Some of the disadvantages are the possible overloading of the element, and the inherently poor sensitivity at the lower frequencies. It is also possible to use the step function response, which, for a linear component is the integral of the impulse response. When the response of an element to an impulse is known it is possible to derive the response to any arbitrary source function.

When the input to an element is random, correlation analyses are useful. The correlation function of the output of the element may be obtained by convoluting the correlation function of its impulse response with the correlation function of the input. A mathematically equivalent expression of this relationship is: the power spectrum of the output of a linear system, in response to a random input is equal to the square of the magnitude of the system function multiplied by the power spectrum of the input function. When the input function consists of "white" noise, that is, uniformly distributed in frequency, the impulse response of the system is the crosscorrelation between the input and output divided by the power spectrum of the noise — which, for white noise, will be a constant.

Primary emphasis thus far has been placed on the interactions between the input signal structure and the linear system as indicated by the structural modifications indicated at the output. Linearity in conjunction with the use of orthogonal components as structural elements facilitates energy conservation criteria and fulfilling boundary conditions. For many important processes, however, minimizing the loss of signal energy may be an inadequate assessment of performance. In simple detection problems it is necessary only to determine the existence of a particular signal, and knowledge of the signal energy and some measure of the interference permits quantitative evaluation of detection performance. If sufficient information is available regarding the interference, then the detection criterion may be extended to include the occurrence of false alarms related to the probability of detection. As was evident from the early sections of the survey, there are many forms which "signals" and "interference" may assume — many methods by which their various characteristics may be described, and there are many tasks which must be performed. In addition to determining the existence of a signal, the spatial location of the target which may be reflecting or radiating signal energy — is required, along with temporal changes of location. Identification may be made in terms of spatial extent, shape, and temporal changes of these characteristics. Interference may originate from within the receiver components, or may be caused by unpredictable fluctuations of propagating conditions within the medium or of the boundaries. When a system is intended to encompass a wide range of dynamic conditions — it is evident that signal-to-noise formulations are inadequate.

Earlier it was indicated that since probabilistic measures were not intimately associated with physical attributes they possessed the quality of extensibility — and could be usefully employed when the problems involved large numbers and multiple variables. Other measures, entropy and information — had the additional virtue of quantitatively interrelating probabilities; these measures being relatively invariant to the type of attribute. Decision theory concepts extend the generalizations further by including as part of the formulation of the problem such features as the use of a priori information, and quantitative assessments of performance and costs. Some of these assessments may be derived by applying signal-to-noise criteria to individual tasks. However, decision theory permits a more concise and complete evaluation since more of the important factors are included as an integral part of the description.

As the problems grow to encompass the greater complexity of space-time processes — information "fields" — some conceptual correspondence still remains between methods employed for the simpler elements, and the methods which are necessary for the more complicated cases. Simplified quantitative measures may be obtained from space-time sampling, space-time correlation functions, and space-time ambiguity functions. Deterministic and probabilistic measures along with entropic and informational measures may be employed and for reasons similar to those given in the earlier sections. A powerful generalization which encompasses many physical processes — acoustical and electromagnetic — is contained in the recognition that finite bounds exist on the total information, information rate, and information "density." Ultimate limits of observation and hence of descriptions, arise from the interplay between information and entropy, between what is known and what is unknown. Physical processes may not be observed and hence may not be represented in infinite detail while still preserving the features upon which the observation rules are based. Although a complete analysis is not always justified, recognition of the basic rules can provide guides leading to a more economical utilization of the available commodities.

This survey represents a first step — an initial probe, and although a large number of facets have been discussed — many others have been omitted. Hopefully, it will serve to stimulate additional effort, and even in its present form it will provide a useful reference framework within which analytical and instrumental concepts of importance in space-time processes may be examined and evaluated.

* * *

www.ingramcontent.com/pod-product-compliance
Lightning Source LLC
Chambersburg PA
CBHW081523220326
41598CB00036B/6315